The
Slate Roof Bible

Everything You Wanted to Know About Slate Roofs
Including How to Keep Them Alive for Centuries

Joseph Jenkins

Text printed on 70 lb. Halopaque, 100% recycled paper, processed chlorine free. Layout done on an IBM compatible 486/120mHz-DX4, with Quark Xpress 3.32 and Photoshop 4.0; output on HP LaserJet 5L. Printed by Gilliland Printing, Arkansas City, Kansas.

The Slate Roof Bible

Everything You Wanted to Know About Slate Roofs Including How to Keep Them Alive for Centuries

© 1997 by Joseph Jenkins - All rights reserved.
ISBN 0-9644258-0-7
Printed in the United States of America
First Printing, August 1997
Published by Jenkins Publishing, P. O. Box 607, Grove City, PA 16127 USA
Distributed by Chelsea Green Publishing, PO Box 428, White River Junction,
VT 05001 (1-800-639-4099)
Library of Congress Catalog Card Number: 97-93324

Comments, inquiries and suggestions for further revisions of this book are wel-come and should be addressed to the author c/o the publisher. Slate roof restora-tion contractors, suppliers, and manufacturers of slate roofing tools, equipment or materials not already mentioned in this book are also encouraged to contact the author via the publisher.

Address book orders, reader feedback, and inquiries to:

Chelsea Green Publishing, PO Box 428, White River Junction, VT 05001
1-800-639-4099 (802-295-6300)

Another title by Joseph Jenkins also available from Chelsea Green:
The Humanure Handbook - A Guide to Composting Human Manure
[(1994), 198 pages, softbound, illustrated, indexed; $19 + $4 s&h]

Ask for either of these titles at your favorite library or bookstore.

*Dedicated to the late Peter Odrey, who got me started on roofs in 1968;
and to my parents, Dorothy D. Graham Jenkins and Joseph J. Jenkins, who got me started.*

Acknowledgements

The author wishes to thank the following persons for aiding him in his search for information: **Kenton Lerch**, Structural Slate Co., Pen Argyl, PA; **Paul Boyce** of Schenectady, NY, Curator of the East Poultney Historical Society, for providing Ruggles Equipment Catalog illustrations shown in Chapter 5; **Clark Hicks**, Evergreen Slate Co., Granville, NY; **Susanne Rappaport**, Slate Valley Museum, Granville, NY; **Don and Ruth Ann Robinson**, Delta, PA (Ruth Ann is curator and President of the Board of the Old Line Museum, Delta, PA); **Noel N. Moebs**, U. S. Dept. of the Interior; **David Crummette**, Buckingham Virginia Slate Corp., Arvonia, VA; **Al Drent**, Operations Manager, Rockmart Slate Corporation, Rockmart, GA; **Dan Morris** (architect), Dolgellau, Wales; **Ann Thomas**, Snowdonia National Park Study Center, Maentwrog, Wales; **Chuck Smid**, New England Slate Company, Sudbury, VT; **Doug Lodge**, Lodge Slate Roof Restoration, Cochranton, PA; **John Neil**, Newfoundland Slate Inc., Ontario, Canada; **David Large**, Newfoundland Slate Inc., Canada; **Barry Smith**, Smith Slate Roof Restoration, Union City, PA; **Doug Raboin**, Twelfth Century Slate Roofing Company, Burlington, MA; plus many others who have helped along the way.

Editorial assistance was generously provided by **Joseph J. Jenkins**, **Jeanine Jenkins**, **Barry Smith**, and **Philip Terman**.

Special thanks to **Dave Starkie**, who lives somewhere in the UK, for many of the European photographs in this book; and to **Tom Griffin** of the Otter Creek Store, Mercer, PA 16137, for the cover art, some line art in Part II, and general design consulting. Back cover photo by Jeanine Jenkins.

Some of the artwork in this book originated in the book *Slate Roofs,* published in 1926 by the National Slate Association, and now in the public domain, although most of that artwork used here was modified from the original drawings, many of which were inaccurate. Additional illustrations from public domain sources are credited. Some illustrations were used from electronic clipart. Many of the references cited at the ends of the chapters are out-of-print and are not notated in the text of this book. Some of the more current references are, however, notated in the text.

The Slate Roof Poem

Philip Terman

Have you seen a slate roof glisten as the sun rounds the ridge?
Have you seen it shimmer in a splash of a sky full of stars?
Have you ever seen a close-up of slates interwoven on a roof like leaves
on a tree or the way slats of stones will seem to quiver in a creekbed,

or how from a distance they can look like steps ascending toward the sky,
or how they float in mid-air, the way they hang loose on their nails,
or how, roofers will tell you, they make a joyful noise in the way they are tapped
and when rain patters like a waterfall as you lie under them listening to their music . . .

Because the rock was here before we came and will be here after we leave.
The earth is the rock that our flesh and bones were formed out of, rock of our ages --
our first dwelling and our last, the veins we tap, the secrets we extract.
Have you driven by a farmhouse and noticed the roof with its year slated in,

or a chicken coop with a heart-shape or a gravestone carved with a poem?
The slate on my study in Pennsylvania is from a farmhouse in New England.
Would we recognize each other, we who were protected by the same substance?
The slate outlives the house under it and the people inside it. Consider

the spaces of time elapsing between when the silt deposited and collected
in the ancient river beds and compacted for millions of years -- shimmying
and hopping and bopping to the geologic boogie -- into what is now your roof
as you sit shaded from the sun or drying under the rain. Consider

all that mineralogical pressure, the way in our flesh and blood, fears
and desires, bits of memory and insights, can crystallize into some force
that is as magnificent as the roofs of spires and cathedrals, how out of our own
crystallized mica we can be a protector and at the same time splendid. Consider

how it must have been created on the first day, with all the other stuff of the earth.
Consider how its origins are a mystery, like love and death and the desire
to use a material that will last beyond out own lifetimes, like planting trees
and making art, impractical and out-of-fashion, we of the long term,

we who have faith in the future, we who play our minor role
in the eternal drama of the significant stone, which will last eternities
without us, we who want to know what our small part of forever feels like.
Because each square of slate is really a petal of a flowering sediment,

a small piece of earth from within the earth. As children we were honored
to be called upon to walk to the large black wall of slate in the front of the room
and write our names and drawings and scribble our confused answers
and we were rightfully reprimanded to stay after and erase them.

A slate roof looks like a book before it is bound. All of the pages
are written by a collaborative effort of weather and time. Hold one slate
in your hands. It is smooth and in places crinkled like handmade paper.
You can't see yourself in the surface, it has lasted and will last much longer

than you will, but it is beautiful, it makes any building -- pigsty or majestic dome --
like a quaint hut, a cottage in a fairy tale. Because it is so of-the-earth
it is otherworldly. Something mysterious about the way it is always shimmering,
something perplexing about the way it seems to absorb all history and stay its place.

One cannot imagine a slate roof without farms or woods or flowers around it,
someone baking bread beneath it in a satisfying silence, or simply staring
out a window as dusk tells its story. In one dream we walk on slate
but we slide because it is slick like trying to stand on water. In another

we are in a burning building but we are safe inside our walls of slate.
Someone gives us slate pencils and tells us to mark our messages. Another
perhaps in the next century will decipher its meanings and what we say
and how we say it will suggest something of our lives and the way we lived them.

Even if they have been effaced, even if our most profound secrets were lost,
something of what we held for a short time will be held by others,
what our hands have smoothed will be smoothed by them and set
carefully to bear in their rightful place on the roof with all the rest, fastened

exactly like our own and offering like our own their significant securities,
making again their music and light. Imagine the sky full of pieces of slate.
It covers the earth the way a roof covers a house the way in some faiths
one covers one's head because we are all in the presence of the holy.

Philip Terman is an Associate Professor of English at Clarion University of Pennsylvania. He is the author of two volumes of poetry -- *What Survives*, and *The House of Sages*; and he lives under a slate roof in Barkeyville, PA.

Contents

Part I - Know the Roof

Part II - Keep the Roof

Death of a Barn

I watched a hundred year old barn burn to the ground one day. As the crackling flames shot up in the air rapidly engulfing the old building, a large raccoon suddenly appeared on the ridge of the roof. The raccoon had been roused from its daytime sleep in the hayloft by the smoke and flames and shouts of the firemen, and had quickly scampered up the oak barn beams to find a safe place to hide. As the flames spread and the heat grew more intense, the raccoon began to suspect that it had made a wrong move and started pacing nervously back and forth along the ridge. Those of us on the ground who were gaping at the spectacle were horrified by the thought that the raccoon was about to become broiled alive, right before our eyes. The fire-fighters tried to knock the animal off the roof using the spray of the firehose, but the raccoon, not as dumb as it appeared, simply dodged the water by ducking behind the roof ridge. Soon the situation grew unbearable for the animal as the flames licked over the barn roof, and the raccoon, in an obvious act of desperation, suddenly made a swan dive off the gable end into a clump of bushes forty feet below. It looked like a suicide dive to me, but the animal surprised everyone by bouncing off the ground with a thud, then running away. Perhaps raccoons, like cats, have nine lives. Not true for barns.

In this case, the owner wanted the old, hand-hewn barn destroyed and had called the fire department to burn it down. Why? It was, after all, a beautiful, century old structure, a memento of the agrarian days of rural Pennsylvania, of a time when everyone out in the country had farm animals and lifestyles centered around animal husbandry. When the barn was built just before the turn of the century, people still got together and helped each other. They didn't hop in a car and drive to each other's farms, because there weren't any cars. There were hardly any roads, and what roads existed tended to be rough and impassable in bad weather. There wasn't any electricity out on the farms either. People relied on their horses for transportation, and if the roads were too muddy for horse and buggy, they just stayed home, or they walked.

A cooperative group had helped raise this barn, probably the same group who got together at the local church on Sunday to pray for good crops, or a safe birth for a young couple, or life renewed for a sick elder. On Saturdays they congregated at the local grange, or in this very barn to plan political strategies or just to have a good time square dancing by the light of kerosene lanterns, if not the moon.

When the barn was built, it symbolized all the hope and promise of a developing country. People were settling the land, clearing the forest and planting the fields. Their survival rested upon the constructive relationships they maintained with their neighbors and their cows, horses, sheep, pigs, goats and chickens. The barn would house these animals, and thereby provide a livelihood for the country people whose new homesteads sprang up on the Pennsylvania country-side like mushrooms after a rain in the years following the Civil War.

The barn bore testimony to the use of local trees for construction, and the beams were wrought from white oak, chestnut, or yellow poplar, while the siding was cut from pine or hemlock. The post and beam frame required no nails, but was fastened together with elegant mortise and tenon joints locked into place with stout wooden pegs called trunnels or "treenails". The beams were shaped by hand with broad axes, adzes, and chisels; cut with hand saws of various types, and drilled with breast drills, braces, bits, and augers. No power tools were used to build this barn, which remained square and erect after a century of use. The proud builders no doubt stood back and admired their work when the construction was finished a hundred years ago, then they communally feasted and together celebrated their accomplishments, their communities, their country, and their lives.

The elderly widow who eventually owned this farm called me one day and asked me if I wanted the roof. The barn had to go, she said, as it was beginning to collapse and was presenting a hazard. I could have the roof if I wanted it, as well as any other parts of the barn, but I only had two weeks to get it. I went over to the farm and inspected the barn to see if salvaging the slate roof was possible. Some barns are so far gone that climbing on the roof would pose an unacceptable risk. This barn only had one corner rotted beyond reasonable repair; the rest of it was in good condition, so I agreed to remove the roof.

Later, when I climbed on the barn roof, I saw why that bad corner had rotted: three slates were missing on a spot that was, coincidentally, directly over a main support post. Aside from those three slates, the rest of the sea green slate roof was in good condition. Of the 3000 original slates on the roof, three of them, or one tenth of one percent, had fallen off and never been replaced. The resulting hole in the roof was on the back of the barn where it was not easily seen, and therefore was easily ignored, and the owners had allowed the rain water to leak through the roof decade after decade until the barn was crippled, then destroyed. Three slates take about one hour to replace, including setting up and taking down ladders. For want of one hour of professional maintenance, another hand-hewn barn, a symbol of America's rural heritage, died.

The barn wasn't a complete loss, though, because the slate roof was salvaged and used on a new building. And I'm willing to bet the clever raccoon found itself a new home too, but probably not in a barn.

▲ Thatch and slate abut in Devon, England
▼ The other side of the roof shown above, detailing overlapping thatch ridge.

(Photos by Dave Starkie)

Chapter One
Why You Should
Keep Your Slate Roof

"Beneath the moss covering the roof of the rugged Saxon Chapel, erected during the eighth century, at Bradford-on-Avon in Wiltshire, England is a slate shingle roof in good condition after twelve centuries of exposure to the elements in one of the most severe climates in the world, a monument to the enduring qualities of slate." H. O. Eisenberg

The telephone rang. I picked it up. "Hello," a lady answered, "Are you the guy who buys used roof slate?"

"Yes."

"Well, I have enough for a whole roof, and you can have it all if you want them."

"Where are they?"

"They're on my house!"

"They're still on your roof?"

"Yes! And you can have them for nothing if you'll take them off."

I paused for a moment, then asked, "Why do you want your slate roof taken off?"

"Because it leaks. We've already bought fiberglass shingles to replace the roof; they're sitting in the driveway. We just need somebody to take the old roof off."

"Well ma'am, the reason I buy roof slate is because I repair and restore slate roofs for a living. Maybe I should have a look at your roof before you take it off."

There was a pause at the other end of the line. I could almost hear the thoughts racing through the lady's head: What? I can *repair* my slate roof?

"But we already bought the shingles."

"You can return them and get your money back *if* you don't need them, which you may not. What *kind* of slates do you have on your roof?

"What kind? I didn't know there were different kinds."

"Slates generally fall into two categories - hard and soft. If you have hard slates, they should last the life of your house and you won't need to replace them. If you have soft slates, then you probably don't have a choice, you'll have to replace the roof. I can tell at a glance what kind of slate you have."

The next day I stopped to look at the roof. The slates were hard Vermont "sea green" slate with a life expectancy of 150 - 200 years. The house was about eighty years old. There was one slate missing from the roof - *one slate!* - and the roof had a little leak at the spot where the slate had broken off. Otherwise, the roof was beautiful. So I offered to repair the roof for a small fee, explaining that the roof should never have to be replaced in her lifetime. The lady accepted, I did the hour-long repair job, she saved both her roof and several thousand dollars, and I haven't seen her since. This is a true and typical story.

I can go on and on with these kinds of stories. One young lady who had a beautiful old Victorian house with a hard, Vermont slate roof in very good condition told me she also was con-

Figure 1.1: This 1860 church roof in Fair Haven, Vermont was 135 years old when photographed, and is still in good condition. The roof is made of Vermont purple slate with a Vermont green slate date.

(Photo by Jeanine Jenkins)

sidering having the slate taken off the roof. "Why?" I asked. "Well I thought you were just *supposed* to replace slate roofs when they got old," she replied. We were standing in her front yard when she said that. The summer sun was glistening off her stone roof. Her impeccable white house stood like a majestic tribute to a time gone by when things were built with quality in mind -- built to last. The roof had been taken care of by her father, a spry eighty-five year old man who knew something of value. Now the daughter owned the house and her first thought was to rip off the slate roof, which didn't even leak. I stared at the house, at the perfectly good hard-slate roof, wondering how many thousands of dollars it would cost her to remove her stone roof and put on a cheap artificial roof. I tried to imagine how asphalt shingles would look on that proud home.

"No, you don't have to replace slate roofs when they get old," I finally said. "If the slate is hard, like yours is, you can expect it to last as long as your house lasts. It will certainly last your lifetime, and probably your children's."

"Really!?"

Recently, a man called me about the roof on his church. He said he had good, hard, black Pennsylvania "Peach Bottom" slate on the church roof, which is some of the best slate in the world, but he thought it should be replaced. Why? Because the valleys and flashings (metal joints) on the roof had deteriorated and were leaking.

"Well, replace the metal!" I advised. But my advice went in one ear and flew out the other

> *"No, you don't have to replace slate roofs when they get old.*
> *If the slate is hard, like yours is, you can expect it to last as long as your house lasts. It will certainly last your lifetime, and probably your children's."*

as the man explained to me how much it would cost to replace the entire roof with new slates.

"We'll have about $120,000 to $140,000 in replacing that roof," he said.

"Don't replace the whole slate roof, just replace what *needs* to be replaced," I replied. "It's not that hard to replace roof *metal*. By keeping the *slate* but replacing the *flashing* you'll cut your costs to a fraction of what you're talking about. Probably down to *less* than 10%. The Peach Bottom slates you have on your church might last four hundred years, maybe longer. Why take them off? It's a lot of work to take off a slate roof, and it doesn't make a whole lot of sense to do so when the slate is still good. Besides, you can't even get Peach Bottom slate any more, it hasn't been quarried in decades, even though it was once internationally recognized as the best slate in the world."

The man was in a hurry, though, and he didn't seem to hear what I was saying. He had to call someone *else* about removing the church roof (since I wasn't interested). End of conversation.

Then there was the guy who had the beautiful Peach Bottom slate roof on his old Victorian house. Peach Bottom slate that's 250 years old is still good, and no one knows how long it will last. Four hundred years is a fair estimate (as will be discussed in Chapter 7). This fellow decided to have a professional roofing contractor look at his 90 year old roof and estimate its remaining life. He was careful to find the oldest, most established roofer in the city. The roofer had only to take one look at the roof to state with certainty the number of years the roof had left: "Ten," he stated with the air and authority of a professional. And of course that declaration was

Figure 1.2:
This roof near Franklin, PA, probably dated 1887, is still very functional after nearly 110 years.
(Photo by the author)

followed by the next, obvious one: "You'll have to tear it off. Would you like me to give you an estimate?" The homeowner was cautious and intelligent enough not to take the contactor up on the offer. Instead he saved his roof by repairing it at a fraction of the cost of replacement, and he'll never have to put another roof on that big old house as long as he lives. Otherwise, if the slate had been removed, he would have been destined to replace the roof about every twenty years at increasing cost. Which, by the way, is what roofing contractors like to see.

Slate roofs originated in Europe, and the earliest Europeans to develop a proficiency in slate quarrying and slate roofing were the Welsh. Slate roofs are common in Europe where they're valued enough that some are still in good repair after several centuries. It is against cultural mores to replace a slate roof with an asphalt roof in Europe, and the people there are so adamant about it that they have passed laws preventing people from replacing slate roofs with cheap substitutes. There are photos of European slate roofs in this book to help illustrate the contrast between American values and European values. There, they know a good thing when they see it. Here, we rip our slate roofs off at the first sign of a leak and replace them with substitutes that are miserable by comparison. If you own a slate roof, you will likely hear every conceivable excuse to tear it off. The strongest urging may come from contractors, who stand to gain financially from the destruction of your stone roof.

I hear it all. One common myth is that slate roofs have to be replaced because their nails go bad. This is not true *(usually).* At the time of this writing, I personally have worked on well over 700 slate roofs over the past 30 years. *Old* slate roofs. That means I've climbed on 700 slate roofs, torn them apart and put them back together (to some extent), examined them, assessed them, removed and recycled them, repaired and restored them. Their average age has been about 100 years. Most of the nails are still good on these roofs. In fact, they're so good that I could use them over again on a new roof. *Even after a century!* And 99% of these nails are hot-dipped galvanized, *not* copper.

In some cases, on some parts of roofs, especially leaking and neglected roofs, nails can and do go bad. This is especially true when the *wrong* nails were used *in the first place*, which happens, but not very often. But as a rule, bad nails are virtually *never* a reason to replace a slate roof. If the nails are going bad, it's either because the slates themselves are going bad, meaning they're old soft slates, or the wrong nails were used when the roof was installed. However, in Europe, when a roof goes bad because of rusting nails or rotting wood underlayment, the slates are simply removed, the wood replaced, and then the same slates can be renailed to the same roof with new nails. A slate roof can be kept alive for centuries in this manner.

Another dubious reason to want to replace a slate roof is because the roof paper (underneath the slate) has deteriorated. After a hundred years, or less, the roofing felt paper turns to dust. People peer through the gaps in the roof boards in their attics and see that the paper has turned to dust and they conclude, *"Time to replace the roof!"* Wrong! Roofing felt paper is not a necessary part of roofs. It is typically used as a temporary cover to protect the building in the event it rains, until the finished roof is installed. Roof paper *does* help temporarily to insulate the roof, but its usefulness in shedding water is greatly reduced when the finished roof is nailed through it (because thousands of holes are poked through the paper when the slates or shingles

> *I've climbed on 700 slate roofs, tore them apart and put them back together, examined them, assessed them, removed or recycled them, repaired or restored them. Their average age has been about 100 years. The nails are still good on these roofs, in fact they're so good that I could use them over again on a new roof. Even after a century! And 99% of these nails are hot-dipped galvanized, not copper.*

are nailed on). Barns typically do not have *any* roof paper under their slates, and they are as waterproof as any other roof.

Then there's the *curtain* theory. One middle-aged couple called me to look at their roof. They were considering removing it and replacing it with fiberglass shingles; you know, the kind that last twenty years then curl up like potato chips and blow away in the wind. Anyway, I stood out in front of the house looking at the beautiful, decoratively-cut Vermont purple slate roof, one of the best hard slate roofs a house can have, with the homeowners standing by my side.

"That's where it leaks," said the lady as she pointed to a section of roof on the front of the house where an inexperienced repair person had spread tar around.

"Not hard to fix," I replied, then I walked around the house to get a good look at the rest of the roof. I could tell that the man wanted to keep the slate roof, but the woman didn't. Every now and then I'd overhear bits of their conversation as they continued to stand out in their front yard gesturing toward the house.

"The slate roof doesn't match our new curtains!"

"I know, dear."

"We need a brown roof if we're going to match the curtains."

"I know, dear."

When I returned to the couple, I briefly explained to them that their purple slates would last the life of their house, and could last hundreds of years, nobody really knows. Purple Vermont slates that have seen a century of wear look pretty much the same as new purple Vermont slates, so who knows how long they'll last. I told them it would cost about $300 to do the routine maintenance on their 90 year old roof, as opposed to $3,000 to replace the slate with asphalt shingles that were *guaranteed to fail* in twenty years. The lady was adamant about not liking the color though, and I sensed that this roof was doomed. "If you do decide to remove the roof," I offered, "I'd be willing to remove the slates free of charge, for salvage purposes."

"We'll call, we'll call," they said. A month or two later I drove by the house. A brand new, *brown* asphalt shingle roof sat on top of the home. I knew what happened to the beautiful slates - they were destroyed by the intellectually-challenged roofers who tore them off. I had seen it many times - the slates are rudely ripped off, thrown to the ground, hauled off in the back of a dump truck, and unceremoniously dumped in a landfill. The homeowners are several thousand dollars poorer, they've permanently lost their slate roof, and now they're faced with having to replace the asphalt roof every twenty years at a constantly increasing cost. But, in this case, the roof matched the curtains! At least I suppose it did - I didn't notice the curtains - who ever does?

Figure 1.3: Old Roofing Material is Filling Up Landfill Space

When old roofing is removed from buildings it is trucked to landfills. Non-biodegradable, petrochemical roofing made of asphalt and fiberglass typically lasts about 20 years, then is discarded. Roofs such as these that are guaranteed (by the manufacturer) to *fail* in 2 decades, are guaranteed to clog landfills year after year. According to scientific studies, construction and demolition debris makes up 28% of the weight and 28% of the volume of the mixed refuse in landfills, and is much more significant to landfill management policy than styrofoam, fast-food packaging, disposable diapers, and the totality of plastic packaging *combined!* About one fifth of all construction debris is roofing scraps.

Environmental concern is only one reason to restore slate roofs, but it's an important one.
(Source: Garbage Magazine, Mar/Apr 1992, p.67 and July/Aug 1992, pp. 20 - 21)

The Burning of "The Mystery House".

Though staged primarily to prove the fire resisting qualities of certain building materials, the "Mystery House" of Indianapolis was an advertising feature that aroused public curiosity, attention and approval.

The house was built by the Williams Creek Development Co. for the purpose of a fire test. The house is a two story building made of fire proof material and roofed with slate.

To provide a severe test, the lower floor was filled with kindling and lumber scraps saturated with kerosene. The oil-saturated wood, highly combustible, was ignited, and for three-quarters of an hour (when the fire was extinguished) the building confined the flames like a huge furnace.

To extinguish the flames, the fire department deluged the building within and without. A careful and critical examination after this severe treatment, showed the slate roof intact without a single slate cracked from the fire and water.

Publicity of the test was heightened by the presence of the Mayor and the City Fire Chief of Indianapolis. The Mayor started the fire and the Chief put it out.

(From a May, 1926 issue of the Slate News Bulletin).

Official inspection of the "Mystery House" after the fire.

Every now and then one sees a hard slate roof where one half of the roof has been replaced with asphalt or fiberglass shingles. Picture this if you can: you're standing on the ridge of the roof and to one side of the ridge you see a slate roof, the other side a shingle roof. The shingle roof is curling up and falling apart, worn thin, brittle and leaking. The slate roof, already 90 years old, looks nearly the same as it did when first installed, and exactly the same as it did when the shingle roof was installed 20 years ago. You wonder why the homeowner ever took the slate off the one side in the first place. Now look into the future: the shingle roof will soon be replaced, the slate roof will, once again, be left alone. In another twenty years the same scene will repeat itself. Then twenty years later the scene will repeat itself again. Then again. Then again. Each time, the shingle roof will need to be *completely* replaced. The old shingles will be hauled to a landfill, *if there is any landfill space left by then*, and the homeowner will fork out more money to roofing contractors.

The slate side? If left alone, and repaired occasionally as needed, it will smugly sit there, stone quiet, stoically oblivious to the vagaries and vicissitudes of the human race. It will continue to do what it's supposed to be doing: protecting the dwelling and sheltering the people who live there. And it will do so with a level of aesthetics and a richness of history that a fiberglass shingle couldn't hold a candle to. The folly of replacing such a roof with a cheap substitute becomes clearly evident when one roof displays both roofing materials at the same time, and their perfor-

DECIDES TOO LATE TO REROOF WITH SLATE

$75,000 FIREPLACE BLAZE

"York, Me., Oct. 1 (A. P.).—The home of John C. Breckenbridge, on the York River, burned tonight with $75,000 loss. Sparks from the fireplace set fire to the roof."

—From clipping Boston Advertiser. Sent us by E. A. Bullard.

A few days later we learned that these people were considering reroofing with slate. Had actually wired Rising & Nelson Slate Company a few days before about it. Their home could have been saved had they considered earlier and put on slate, the sheltering stone, before they started up the first fire of the Fall in the fireplace. Buildings far away from community fire apparatus need protection of household fire extinguishers and slate roofs.

Capitalize such cases when you see them in the papers to hasten decisions of those in your city whose homes need to be reroofed to avoid such fire hazards.

W. L. Hassenplug, Philadelphia office, Rising & Nelson Slate Company, recalls a similar case in Maryland three years ago. A home was destroyed from sparks from fireplace chimney. It was rebuilt on original foundations, but first thing the owner insisted on was a slate roof.

From the *Slate News Bulletin*, May 1926

mance can be seen and compared side by side.

A common reason why people feel they need to remove their slate roofs is because *"We can't find anyone to fix our roof."* If they do find someone, they run the risk of hiring a contractor who wants to "gouge" them (i.e. charge a small fortune for repairing the roof). I've talked to various contractors who charge anywhere from five to thirty dollars a slate for individual slate replacement (1996). At $30 a slate, a homeowner will quickly go broke. Quite a few slate roof repair contractors don't specialize in slate and when they're not building decks or installing bathrooms they'll "repair" your roof. Most of the contractors I've met who repair slate roofs do it improperly. And because they're not equipped and experienced to work on high, steep roofs, they charge an exorbitant fee, knowing that if they get paid enough, they'll do the work. So *if* a homeowner can find someone to fix the slate roof, and *if* s/he can afford him, then s/he will probably still have to pay *to have the job done wrong!* Which means they'll have to pay again to eventually have the job done right, and re-doing someone else's lousy work is always harder than doing the job right in the first place. This makes for a frustrating situation for owners of slate roofs, and it's one of the main reasons this book was written.

I recently met a contractor who claimed to specialize in slate roof repair, and who told me he goes through about 15,000 pieces of slate every year doing roof repairs. Then he proudly informed me with a grin and a nod, *"I face-nail every one of 'em."* What roof owners don't know is that every face-nailed slate (nailed through the face of the slate so that the nail-head remains exposed to the weather after the repair is completed) is an improperly installed slate, one that will leak, and one that has been ruined by the roofer who face-nailed it. So here's a guy who's roofing company face-nails fifteen thousand slates every year. Every single one of those slates will have to be taken off and replaced someday. Before they do though, they'll probably leak. This guy gets paid good money to slowly ruin people's slate roofs. And I'm not singling him out intentionally. *He's typical!*

Why do contractors face-nail slates? Because it's a little bit easier and quicker than doing it the right way. Not much, but just a little bit, and some contractors focus more on how much money they're making than on how well they're doing the job. Now there *are* instances when face-nailing *is* appropriate, but they're few and far between. We'll get into all of that later when we cover slate roof repairs step-by-step in the second half of this book.

If you don't want to fix your roof yourself, but do want to hire a competent contractor, make sure the contractor you hire is either experienced with slate or has referred to this book *before* he's set a foot on your roof! If your contractor tries to tell you he already knows everything there is to know about repairing slate roofs (a common line), then ask him a few questions to see whether he's telling the truth. Ask him where your slates came from and how long they'll last; ask him to explain how he replaces a single slate, ask to see his slate cutter, slate hammer, slate ripper, and ladder hooks. You may know a lot more about your slate roof than he does by the time you've finished this book, and if the guy is an imposter, you'll know that too!

There are numerous reasons why old slate roofs should be preserved, in addition to longevity. One is *aesthetics.* In this plastic world a stone roof is a symbol of durability, of quality and craftsmanship. Old slate roofs are *antique* roofs, plain and simple. They're rich in history. Imagine the human effort and struggle that enabled people to bring tons and tons of massively heavy stones up from hundreds of feet underground, split them into shingles by hand, and do it all in the horse and buggy days, before electricity was available. People will give their eye teeth for a hundred year old chest of drawers, yet they'll discard at whim a perfectly useful, irreplaceable, antique stone roof worth $10,000, or much more. Why?

Let's also not overlook the fact that the purpose of a roof is protection. Stone roofs are *fire proof,* and withstand the elements like no other roof. If you live under a properly maintained stone roof then you know exactly what I mean. There's no feeling like the feeling of security you get when the wind is howling and torrential rain or sleet or ice is beating down upon the earth and pummeling one's house, and a stone roof stands guard overhead deflecting mother nature's blows. Living under a stone roof is kind of like living in a cave. It's almost a primal experience, and with a little imagination we can see why such a roofing material appeals to us instinctively. There is no shelter quite as secure from the harsh elements as a cave, and any roof made of pure rock comes close to providing the same level of protection as the earth itself.

Speaking of caves, it's time we took a look at the *Neanderthal Syndrome.*

THE NEANDERTHAL SYNDROME

The Neanderthal Syndrome is an observable roof condition discovered by the author, who has had the opportunity to study the traces of human impact left on slate roofs over the years, sort of like an anthropologist studying the old camp sites of cavepeople.

Many, times I have positioned myself atop a slate roof and marveled at the utter and absolute folly with which the roof had been treated. I have shaken my head in disbelief, theorizing in vain as to what could possibly have caused the prior "roofers" to put replacement slates in

Need to hire a **ROOFING CONTRACTOR?** Look for these

WARNING SIGNS

diches dugg
slat ruffs fixd
kars warshed
will werk fer fud

Questionable
credentials

Receding
Forehead

Arms hang
down to
knees

Hairy
knuckles

Suspicious
looking
tools

upside down and/or backwards, to tar an entire roof to cover a pinhole, to drive steel spikes through the slate roof willy-nilly, like a rampaging lunatic who just escaped from an insane asylum and decided, on a whim, to climb up on someone's roof and start pounding nails in it. No doubt I have seen every imaginable assault and insult to a slate roof, much of which is beyond human comprehension, without any reasonable explanation, and, at times, infuriating enough to become a laughing matter simply for the sake of sanity (mine).

The explanation, I've decided, is simple: Neanderthals never became extinct, they simply evolved into roofers. The traces of their work are on almost every old slate roof I have ever seen. I call these traces "Neanderthal tracks," and I've managed to formulate a rather detailed description of some of the prehistoric individuals who left their tracks on slate roofs.

First, there was Jacko, the one-armed, blind roofer. The slates he used to replace broken slates didn't match the roof at all. When he fixed a nice green roof he used black slate. When he fixed a black roof, he used green slate. When he replaced square slate, he used slates with cut corners. Nothing matched and it all looked bad when he was done, but it didn't bother Jacko - so obviously, he must have been blind!

Jacko had a crew who each had his own style of roof "repair." This crew must have worked really hard for many years, because their work can be found almost anywhere, on thousands of roofs. One of the hardest workers was "Spike." His tactics were simple, (of course) - he kept his nail bag full of 3 & 1/2 inch long *spikes*, and he drove a spike into the roof wherever a slate looked loose. Sometimes he'd drive a spike into the roof for no apparent reason, perhaps just to keep busy when Jacko was around. Of course, a sixteen penny nail driven into a slate roof

Figure 1.4:
This purple and green Vermont slate roof dated 1862 will last for generations if properly cared for despite its age of 131 years at the time of the photo. Age does not destroy hard slate roofs as quickly as improper repairwork.
[Photo by author]

cannot be removed without breaking a few slates in the process, but if it's left in the roof it will leak.

That's where Tarbaby Teddy came in. Tarbaby Teddy loved tar, and he used it for all his slate roof repairs. He kept a bucket tied to his waist and he spread it liberally about the roof. If he was in a really good mood, he'd use a thin tar he could brush on and then he'd really go to town. If a roof had a tiny hole in it, Teddy would tar the whole thing. No pinhole was going to get the better of Tarbaby Teddy!

Jacko, Spike, and Tarbaby Teddy worked on slate roofs for decades, and it's speculated that they grew old and died, but I'm not sure about that. They might be out there still working away, reincarnated as younger Neanderthals. Ironically, these guys are slowly putting themselves out of business, because they deface and destroy slate roofs to the extent that the roof owners can't stand their own roofs anymore. When a slate roof has so many metal patches, tar blotches, and leaks that never stop, it looks ugly, gets frustrating, and drives one to throw one's hands into the air and shout, "Tear the damn thing off!"

Fortunately, virtually any Neanderthal track can be erased, no matter how bad or how

Figure 1.5:
A slate covered church spire in Madrid, at the Plaza Major. [Photo by Dave Starkie]

hopeless the roof looks to the roof owner.

There is one exception to this rule. Old slate roofs that are made of *soft (S2) slate* cannot be restored. Soft slate will flake, crumble and fall apart somewhere between 55 and 125 years. If the slate is hard, the roof can usually be restored and may last for generations. If the slate is soft, an old roof may still last for decades.

Unquestionably, historic slate roofs can have an air of mystery about them. The men who installed them have died and the old roofs have been virtually ignored by everyone since. The origins of these roofs and their remaining longevity and value are difficult to determine for laypeople and professionals alike.

The author of this book, a professional slate roofer, set out on a quest determined to clear up this confusion. In the five-year process, he located the original sources of the various old roof slates, and then visited the actual quarry sites, which range from Georgia, up the eastern seaboard of the USA, through Newfoundland and across the Atlantic Ocean to Wales.

The author was like a bloodhound on a trail, discovering much along the way. Old roofs at each quarry site were examined to determine their conditions after a century or more of wear, in order to help ascertain the longevity of the slates from each region. Museums and libraries were scoured for historical information in order to understand the conditions and techniques involved in producing the slates still gracing older homes and buildings today. Personal interviews with quarry workers, roofers, architects, homeowners and other history buffs, along with a multitude of photos, topped off the investigation. The results are revealing, and described in the following chapters.

Figure 1.6:
European culture, perhaps because it is much older than American culture, seems to take the preservation of its slate roofs much more seriously than Americans. This slate dome in Madrid, Spain illustrates the level of roofing craftsmanship that has been achieved in Europe.

[Photo by Dave Starkie]

Figure 1.7

▲ This roof in central Yorkshire, England, shows slates of random sizes, with two pieces of glass for skylights worked into the roof like shingles.

▼ A decorative slate roof in western Wales.

[Photos by Dave Starkie]

Figure 1.8
▲ A Delabole slate roof in Tintagel, Cornwall, England.
▼ A close-up shot of the porch showing the random artistry of the stonework.

[Photos by Dave Starkie]

▲ Slate roof in Crediton, Devon, England, showing ornate ceramic ridges.
▼ Old church in Herefordshire, England, on the Welsh/English border showing Welsh slate.

[Photos by Dave Starkie]

Chapter Two
From Rock to Roof
(What is Slate?)

"Slate tombs high in the Alps near Oisans, France (which, from money and jewels found in them, archeologists have concluded were constructed about 500 B.C.), are still in good condition."
Oliver Bowles

It will help to understand slate roofs if one first understands what slate *is*. Yes, it is a rock, it comes from the ground, sometimes it's black, sometimes it isn't. Most people, believe it or not, don't even know that much. A newspaper reporter once did a story on a local slate roof restoration business, and the first thing she asked was, "What's a slate roof?" The roofer replied, "You are joking, aren't you?" and she said, "No, I've never seen a slate roof. I didn't know there was such a thing." "Next time you drive through town look up," he said. "The slate roofs are the ones made of stone. They usually have a sheen to them, they look natural, they tend to be on older homes, you can't miss them. There are thousands of them right in our small town, and millions throughout the United States and the world."

For those of you who want to know more than that about slate roofs, let's start with some fundamentals. There are three kinds of slate: mica, clay, and igneous. Mica slate is the only kind we're concerned with because that's what roof slates are made from.

Mica slate is considered to have formed from clay-containing silts originally deposited under water in horizontal beds, such as on the floor of ancient river beds or seas. These clay sediments compacted over many millions of years under the pressure of sedimentary deposits above them so that the final clay content became very small through metamorphosis. Some beds were subjected to horizontal geologic forces which forced the beds to fold, or heave vertically, or even turn upside down in the earth.

This geologic pushing around of the clay deposits forced the material to undergo fundamental changes in its chemical composition, eventually to become what we now call slate, made up primarily of mica in the form of fine flakes arranged in parallel order. *Mica* is a generic term

Figure 2.1:	**MINERAL COMPOSITION OF AVERAGE SLATE**

Quartz	31-45 %
Mica (sericite)	38-40 %
Chlorite	6-18 %
Hematite	3-6 %
Rutile	1-1.5 %

Mica is here also known as secondary muscovite, or white mica, chemically composed of potash and aluminum. Chlorite is a mica-like mineral usually containing aluminum, iron and/or magnesium.

[Source: Bowles, The Stone Industries, 1934]

The Most Boring Definition of Slate Imaginable

The American Standards of Testing and Materials has adopted the following formal definition of slate:

Slate is a micro granular crystalline stone derived from argillaceous sediments by regional metamorphism, and characterized by a perfect cleavage entirely independent of original bedding, which cleavage has been induced by pressure within the earth.

for any group of minerals that crystallizes in thin, easily separated layers. In slate, the mineral is primarily silicon dioxide, often in the form of crystalline quartz.

The age of slate ranges from *Cambrian* (from the word *Cambria,* which is a name for Wales) a time 600 million years ago when life dawned on earth and the first abundant marine life appeared, to *Silurian* (another Welsh name), a time 425 million years ago when mountains were forming in Europe and the first small land plants appeared.

Since slate was originally formed by layers of sedimentary clay silt, the finished product is a finely layered stone that can be easily split, somewhat like a deck of cards. If you lay a deck of cards on a table, then "cut the deck," it's obvious that it can be split in two horizontal halves quite readily. However, if you try to split the deck in two halves *vertically,* forget it. The horizontal layering of the deck lengthwise, or of the slate, constitutes the *cleavage* plane of the slate, as it is the plane on which the slate readily splits. This plane is determined *not* by the sedimentary layering of the slate over the eons as one would suspect (that's the *bedding* plane), but by the geological forces that squeezed the clay deposits together. The cleavage plane may be entirely independent of the bedding plane, which is, by definition, a distinct peculiarity of slate. Some roof slates have darker or lighter bands, known as *ribbons,* running across their face showing where the bedding plane was intersected when the slate was split along the cleavage plane.

Slate can also be split along the *grain,* on a plane perpendicular to the cleavage plane (see Figures 2.3 and 2.4). This "grain" of the slate is very important when quarrying, as roofing slate is usually split so that the longer sides of the slate are in the direction of the grain. This helps to reduce breakage. Or, as one writer puts it, *"In splitting and dressing the roofing slate, it is always done so that the grain runs parallel to the longer side of the rectangle. This grain, although never so marked as that in the timber, has a similar effect upon the strength in different directions* [Peach Bottom Roofing Slate, 1898]."

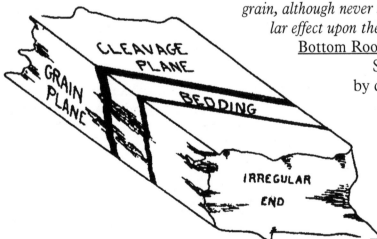

Slate can also be split across the grain, by drilling a hole in the block of slate and wedging a "plug and feathers" into the hole, forcing the block to pop in two (don't try this on a deck of cards).

All this, obviously, can be quite confusing to someone not familiar with quarry geology. If you would like your eyeballs to glass over further with boredom and incomprehension, then read the following industry definitions:

Figure 2.3
A block of slate showing bedding, cleavage, and grain planes.

[From Slate in Pennsylvania, p. 29]

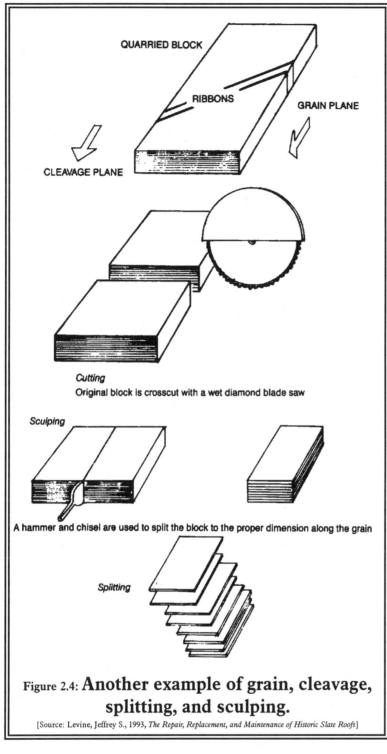

CLEAVAGE PLANE

QUARRIED BLOCK

RIBBONS

GRAIN PLANE

Cutting
Original block is crosscut with a wet diamond blade saw

Sculping

A hammer and chisel are used to split the block to the proper dimension along the grain

Splitting

Figure 2.4: Another example of grain, cleavage, splitting, and sculping.

[Source: Levine, Jeffrey S., 1993, *The Repair, Replacement, and Maintenance of Historic Slate Roofs*]

"The term 'cleavage' has been applied to three different structures in slate. It has been used to designate parting along numerous, close-spaced and parallel planes that bear no relation to bedding; this is true 'cleavage' and the use of the term should be restricted to this feature. The term has also been applied to a tendency to part along small, joint-like openings, more closely spaced than joints, but less so than true cleavage planes, and generally inclined to the cleavage; this may be called 'false cleavage'. A third set of fracture planes of which no trace is visible to the naked eye until the slate is actually broken and which is generally at right angles to the true cleavage, is the 'grain', 'sculp', or 'scallop' of quarrymen. The cleavage plane in slate is the direction of easiest parting and is due chiefly to the arrangement of the individual crystals in the rock." [From Slate in Pennsylvania, p. 29]

I vote we leave the splitting of the slate to the professional quarrymen. Generally, though, you get the picture (I hope).

Slate is a very dense, heavy, finely grained rock, with an average particle size of 0.1 to 0.01 millimeters. This fine particle size aids slate in its ability to be perfectly split, and this cleavability of slate gives it its practical value. A true slate can be split into thin sheets with smooth, even surfaces. Some slates can be split as thin as one thirty-second of an inch. In the manufacture of blackboard slate, a 4'x6' piece can be readily split to a uniform thickness of three-eighths inch or one-half inch.

A scientific theory first published in 1912 by German meteorologist Alfred Wegener stated that all of the earth's land masses were once joined together as a single land mass, called *"Pangaea."* Pangaea is theorized to have split apart approximately 300 million years ago into two continents, one of which eventually became Africa and Europe, and one of which became the Americas and Asia. The basis for this theory is apparent when one looks at maps of the continents today and sees how they seem to fit together, as if they've broken apart from each other. Interestingly, the slate deposits that run up the east coast of the US, across the eastern edge of

HOW OLD IS SLATE?

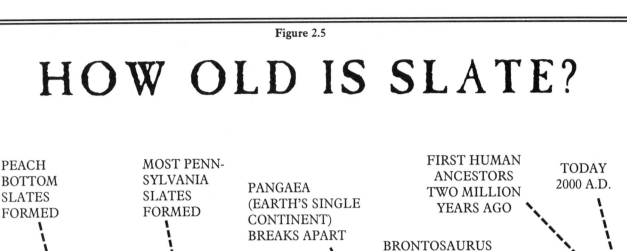

PEACH BOTTOM SLATES FORMED

MOST PENN-SYLVANIA SLATES FORMED

PANGAEA (EARTH'S SINGLE CONTINENT) BREAKS APART

FIRST HUMAN ANCESTORS TWO MILLION YEARS AGO

TODAY 2000 A.D.

BRONTOSAURUS

TIME LINE

700   600   500   400   300   200   100 MILLION YEARS AGO

ERA OF SLATE FORMATION

Slate is so old it boggles the mind.

Humans weren't even a gleam in Mother Nature's eye when the minerals that make up slate were deposited in thin layers by settling to the bottom of ancient seas or river beds. Imagine that you're standing at the end of a time line and the time line extends behind you like a string. Imagine that 10,000 years passes in one foot of the string's length, so that a mere 12 inches behind you is a point in time 10,000 years in the past. If you go back 60 feet, you find North America covered by glaciers. If you go back a mile, you find the earth's major mountain ranges, the Rockies, Andes, Alps and Himalayas, forming. If you go back five miles, you find Pangaea, the earth's single continent, breaking apart. You will have to go back almost twelve miles to find slate forming. If this seems like an unbelievably long time ago, imagine this: the known age of the earth would take us back 85 miles!

Slate predates the Brontosaurus by several hundred million years!

Figure 2.6: # SLATE DEPOSITS and CONTINENTAL DRIFT

The slate found from the southeastern United States, up the eastern US seaboard through New England, eastern Canada, and in Wales and Europe formed in the Earth before the continents broke apart. It is these slate deposits that have been worked into roofing material for American homes. However, slate has also been imported into the U.S. from places such as Austria, Belgium, Czechoslovakia, Germany, Gibraltar, Japan, Netherlands, Sweden, India, Switzerland, Italy, France, Norway, China, Portugal, Australia, Luxemberg, South Africa, Spain, England, and Bali.

[Listing of imports from "Slate in 1929", by Bowles and Coons, Mineral Resources, 1929, Part II, Page 166, U.S. Dept. of Commerce, Bureau of Mines,]

Canada and into Wales evidently formed before Pangaea broke apart, when Wales and Canada were still joined. This is why Canadian and Welsh slate are nearly chemically identical, despite the fact that they're separated by the Atlantic Ocean. They're also both very similar to Vermont slate.

Since slate formed from sedimentary deposits, an ancient river or an inland sea must have existed on Pangaea where we now have slate beds. When one considers that the slate deposits in eastern Pennsylvania are about a half mile thick, and that slate is an intensely compressed material, then one can speculate that an immensely deep body of water must have existed for eons and collected silt for an eternity of the earth's time. This scenario becomes even more fascinating when one understands that during the time of earlier slate formation, no animal or plant life existed on Earth, as the planet was too young, and life in the seas was just beginning to blossom. This may explain why fossils are only very rarely found in slate deposits.

Due to unknown reasons, the slate that formed on Pangaea developed different characteristics depending on the location of the slate bed. For example, the slate in Wales and in Newfoundland, Canada has many color varieties, including black, purple, blue and green. Continue south into Maine and you find only a black slate, but carry on a little farther into Vermont and New York and the slate is either purple, green, gray, red or black, and may be

Figure 2.7: Chemical Composition of Slate (%)

Mineral	Vermont	NY	PA	VA
Quartz	59-68	56-68	55-65	54-62
Al_2O_3	14-19	10-13	15-22	17-25
Fe_2O_3	0.8-5.2	1.5-5.6	1.4-4.5	7.0-7.
FeO	2.5-6.8	1.2-3.8	2.3-9.0	
CaO	0.3-2.2	0.1-5.1	0.2-4.2	0.4-1.9
MgO	2.2-3.4	3.2-6.4	1.5-3.8	1.5-3.9
K_2O	3.5-5.5	2.8-4.4	1.1-3.7	
Na_2O	1.1-1.9	0.2-0.8	0.5-3.5	
CO_2	0.1-3.0	0-7.4	1.6-3.7	0.2-2.0

From US Dept. of Commerce, Bureau of Standards, Journal of Research, V9, No.3, 9/32

streaked, mottled, or layered with iron-containing minerals that change color with exposure to the weather. Travel a bit further south again into eastern Pennsylvania and you find huge deposits of black slate, but a few miles further south in Chapman, PA, the black slate contains grayish bands. Down in Maryland you find both a black and a purple slate that were once quarried there, but continue into Virginia and the slate is black again, and on close examination contains a sparkly quartz that glistens in the sunlight. Black, red and green slate deposits exist in Arkansas; both green slate and black slate can be found in Georgia, and green slate in Tennessee.

The period of slate formation ranged from 600 million years ago or more, to approximately 450 million years ago, which allows for a period of about 150 million years in which slate formation occurred. One hundred fifty million years is too long to comprehend, but if sedimentary deposits occurred over millions of years, one can understand why slate would vary not only according to the location of the slate in the earth's sub-surface, but also according to the depth of the slate in the thick layers of the beds. In eastern Pennsylvania the main commercial slate deposits are 2800 feet thick, thirty miles long and two to five miles wide. That's a big chunk of solid slate lying just below the surface of the earth. Twenty eight hundred feet thick means more than a half mile *thick* layer of slate, and in that half mile are billions of thin layers, each repre-

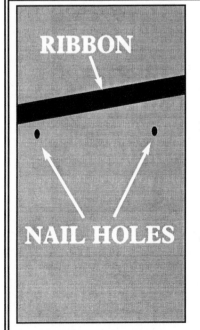

RIBBON

NAIL HOLES

Figure 2.8: Pennsylvania Ribbon Slate

A ribbon can be a band containing a high amount of carbon. If so, the ribbon will deteriorate much more quickly than the rest of the slate, causing the piece of roof slate to eventually break into two pieces. This diagram is showing the ribbon crossing the slate above the nail holes. When the ribbon deteriorates, the slate will not fall off the roof because the nails will hold it in place. The slate may still leak though, as the soft, crumbly ribbon will allow a place for water to penetrate. If the ribbon is below the nails, the slate will eventually break in two and the bottom will fall out. The ribbon may be difficult to see unless one knows what to look for. Many soft, black PA slate roofs are made of ribbon slate, and can't be restored. A ribbon slate may last 50 to 100 or more years, depending on their place of origin. Some slates, such as Chapman slates, have hard ribbons that do not deteriorate prematurely.

senting a period of time so far in the distant past as to be quite unfathomable to us short-lived humans. So as quarrymen dig down through the slate, they come upon layer after layer with different characteristics, some softer, some harder, some junk.

For example, in the Pen Argyl region of eastern Pennsylvania, one of the world's most prolific sources of black slate, the layering of the slate is particularly evident. Beginning with the topmost beds, slate *runs* appear in the following succession: *Pennsylvania Run, United States Run, Diamond Run, Albion Run, Acme Run, and Phoenix Run,* each run bearing a slate of somewhat different quality. The runs aren't in direct contact with each other but are separated by intervening beds 75 to 280 feet thick of unworkable slate-like rock.

Each run is itself then separated into individual *beds.* The Albion Run, for example, consists of 12 beds combined together to form a total run thickness of 184 feet. Some of the beds are "big beds," some are "ribbon" slate (which we discuss in the next paragraph and in Figure 2.8), and some are unworkable rock. The Albion Gray bed is known for its exceptionally high quality.

Some slate deposits have thin (one or two inch) layers of minerals that will leave a visible strip across the finished roof slate. You may see green streaks across purple slate, and various colored streaks or mottled spots on different slate varieties. Normally, these do no harm but instead add some color and character to the slate. However, the most notorious "ribbons" occur in black Pennsylvania slate, where an almost invisible band of gray/black material, high in carbon content, leaves a vulnerable spot in the slate. As carbon deteriorates more readily than the more common slate minerals such as quartz and mica, these ribbons will turn soft over time and the slate will then break apart. A carbon ribbon such as is common in PA "ribbon slate" will turn soft enough over time (50 - 100 years) that you will be able to poke your finger through it.

Other slate, most notably the Vermont "sea green" that is so common in the U. S. northeast, formed from sediment that included layers containing iron. The iron is not visible when the slate is first dug from the ground and the slate appears a beautiful blue-green color reminiscent of the sea. But when the slate is split, any layer left exposed on the surface of the slate

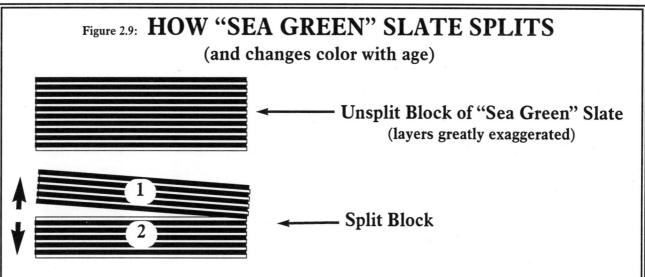

Figure 2.9: HOW "SEA GREEN" SLATE SPLITS
(and changes color with age)

Unsplit Block of "Sea Green" Slate
(layers greatly exaggerated)

Split Block

The top diagram represents a piece of sea green slate showing the layers greatly exaggerated. The bottom diagram shows the chunk of slate split in two as if someone had just tapped the left side of the slate block with a hammer and splitting chisel. The layers of slate are shown in different shades, indicating that some of the layers contain iron (dark layers), and the other layers don't. When the slate is split with iron-bearing layers exposed on the surface, as piece #1 is, that slate will change color and gain a reddish hue on a roof. Piece #2 will not change color but will remain "green" since an iron bearing layer is not exposed. Green layers turn light gray with age, therefore, sea green slate roofs show a mix of reddish-brown and light gray slates.

Figure 2.10
The above Peach Bottom slate gravestone, a scant one inch thick, is dated 1743, and the inscription does not appear worn at all after 250 years exposed to the weather. It reads, "Here lys y body of Jamer Roger who departed this life Novembry 9, 1743 aged 33 years." This stone bears testimony to the longevity of slate from the Peach Bottom, Pennsylvania slate region.

(Photo taken by the author at the Chestnut Level Cemetery, Quarryville, PA)

which contains iron will oxidize, or maybe we should say "rust," and turn a reddish-brown hue. Because not all layers contain the iron, not all slates will turn red; only those slates that have an exposed iron layer on the surface will change color. The result is an interesting mottled red and gray appearance which the roof develops over the years (the green layers turn gray over time). Many, many people have sea green slates on their roofs, as the Vermont sea green slate deposits were abundant, easy to work, high quality, and therefore popular for roofing.

Slates that are harder to work because, for example, the grain is not straight or uniform enough or the slate is too hard and brittle, will not be so widespread on roofs. The same goes for slate that is found only in small deposits - not very abundant on roofs either.

Many people say, "I didn't realize that there were different kinds of slate. I thought all slate was the same!" As you can now see, all slate is not the same. In our younger years, the only slate we may have been aware of was the slate blackboard, and unless we grew up in a slate quarrying area, we developed the idea that all slate is black. Instead, we actually have a great variety of slate with a great variety of qualities, colors, and characteristics. Roofs have been covered with all types of slate, as well as other stone. Some black slate from Italy even turns completely *white* with exposure to the weather!

One of the most important things to know about slate roofs is what *type* of slate is on the

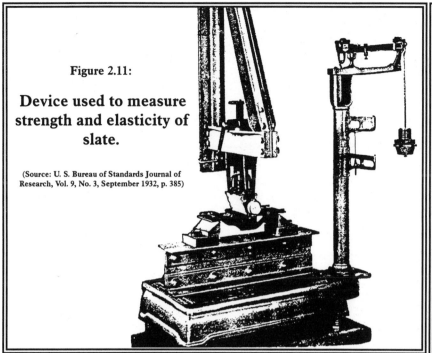

Figure 2.11:

Device used to measure strength and elasticity of slate.

(Source: U. S. Bureau of Standards Journal of Research, Vol. 9, No. 3, September 1932, p. 385)

Figure 2.12

Average Absorption of Slate
After 48 Hours Immersion in Water

Monson, Maine:	0.05%
Vermont-NewYork:	0.13%
Buckingham, VA:	0.06%
Eastern Pennsylvania	
Bangor:	0.28%
Pen Argyl:	0.30%
Hard Vein:	0.16%
Wind Gap:	0.38%
Slatington:	0.29%

The top three varieties of roof slate have a much lower absorption factor and have also proven to last much longer on roofs than the slates listed below that have high absorption factors. This indicates that the less water a slate will absorb, the longer it will last.

(Source: Bureau of Standards Journal of Research, Vol. 9, No. 3, p. 397, September 1932)

roof, because then we will know how long the roof *should* last. We'll look at how to identify the slate on individual roofs in the next chapter of this book.

DETERMINING THE QUALITY OF ROOF SLATE

We have established that some slate is hard, some soft, depending on such factors as geological age, the degree of metamorphosis and the mineral composition. For the sake of simplicity, **soft slate** is here being defined as *slate that more readily softens with prolonged exposure to the weather*, while **hard slate** is *slate that resists softening and stays hard with prolonged exposure to the weather*. Hard slate will obviously outlast soft slate on a roof.

I have seen soft slate (Pennsylvania ribbon slate) that had to be replaced in 55 years, and I have seen hard slate 250 years old (Pennsylvania Peach Bottom slate) that had no apparent deterioration (see photo of slate gravestone, page 28). I keep a piece of hard slate in my shop that I brought back from a 16th century abbey in Wales, an original slate, which shows virtually no wear. It had been hung on the roof by a single wooden peg, two inches long and the thickness of a pencil. The peg had been driven through a hole in the slate and hooked on a thin piece of roof lath, which was the way Welsh slate was originally installed (see photo, page 189).

Measurements of slates from various regions of the United States show that the strength of a slate when newly quarried is not necessarily an indication of how long it will last on a roof. Some slates that last much longer than others will not measure much stronger than the others when newly quarried, and may even measure weaker. After many years of exposure to the seasons, some slate will deteriorate more rapidly than other slate, regardless of how strong the slate is when new. So the longevity of slate cannot be judged by the apparent hardness of the slate

RATING THE DURABILITY OF ROOF SLATE

The American Standards of Testing and Materials (ASTM), first organized in 1898, is a non-profit organization which writes standards for materials, products, systems and services. ASTM C 406 provides standards for roofing slate, distinguishing between S1, S2 and S3 slate according to the criteria below. S1 slate rates a service life in excess of 75 years, S2 is rated for 40-75 years, and S3 is rated 20-40 years. In reality, hard slate (S1 slate) may last 400 years, while S2 (soft) slate may last considerably longer than 75 years.

	Modulus of Rupture Across the Grain, min. PSI	Maximum Absorption %	Maximum Depth of Softening inches (mm)
Grade S1	9,000	0.25	0.002 (0.05)
Grade S2	8,000	0.35	0.008 (0.20)
Grade S3	7,000	0.45	0.014 (0.36)

when newly quarried, and instead, a variety of factors are considered.

The slate industry uses several highly technical standards to determine the strength, elasticity (bendability), abrasive hardness, toughness, acid resistance, and porosity of slate. A variety of contraptions may be used to determine these measurements (see Figure 2.11), thereby allowing for the slate to be categorized according to *expected* durability.

One indication of the potential durability of slate on a roof is the absorption characteristics of the slate. Some slate absorbs more water than others, although no slate absorbs very much. Figure 2.12 shows some average absorption levels for various slate after immersion in water for 48 hours. Those slates that have a very low absorption ability have a very high rooftop longevity.

To simplify the complicated process of rating the durability of slate, the industry has adopted a rating scale which rates slate as S1, S2, and S3. This rating scale combines various factors pertinent to the longevity of the slate such as acid resistance, strength, and ability to absorb water. Slates that have a high acid resistance, low water absorption, and high strength are listed as S1 slate, which is the best slate to use for roofing. Generally speaking, these are the "hard" slate to which we have been referring - the slate that will last several human lifetimes, and perhaps several centuries, on a roof.

S2 slates are not as acid resistant as S1 slate, and they absorb slightly more water than S1 slate, and although they are expected to not last as long on a roof as S1 slate, they still will make an excellent roofing material. Much of the S2 slate found on roofs today originated in eastern PA in Lehigh and Northampton Counties, and appear as black slate, or brownish-black slate (also called "blue-black", and gray). These are the most common "soft" slates in the United States, and they may last only as long as 50 years, or maybe as long as 150 years, depending on the exact place of origin. Many original soft slate roofs have already deteriorated beyond repair and have been replaced, and many more will be replaced in the next few decades. Although soft slates don't last as long as hard slates, they'll still outlast just about anything else on the market.

Soft slates appear crumbly or flaky when old and will give a dull thud when tapped on with a hammer, while hard slates will appear smooth and will ring when tapped. But *some* hard slates *will* appear a little flaky too after a century or so, particularly some varieties of sea green slate. In addition, although *many* black slates *are* soft, some of the most durable roof slates in the world are black. So it is important to try to pinpoint the exact place of origin of the slates on old roofs in order to understand their qualities and to judge the remaining life expectancy of the roof. The next chapter may be of some help.

> *One of the most important things to know about slate roofs is what type of slate is on the roof, because then we will know how long the roof should last. We learn how to identify the slate on individual roofs in the next chapter of this book.*

▲ **Closeup of a slate roof in the Lake District of northwestern England. Note how the slates are random widths, uncommon for old slate roofs in the USA, which are generally of uniform width.** (Photo by Dave Starkie)

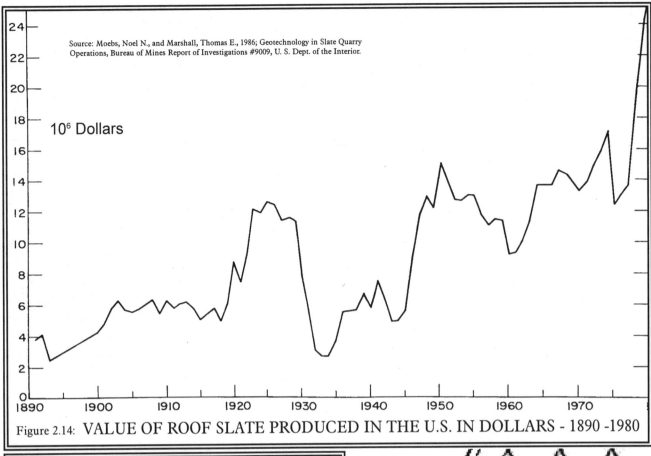

Source: Moebs, Noel N., and Marshall, Thomas E., 1986; Geotechnology in Slate Quarry Operations, Bureau of Mines Report of Investigations #9009, U. S. Dept. of the Interior.

10⁶ Dollars

Figure 2.14: **VALUE OF ROOF SLATE PRODUCED IN THE U.S. IN DOLLARS - 1890 -1980**

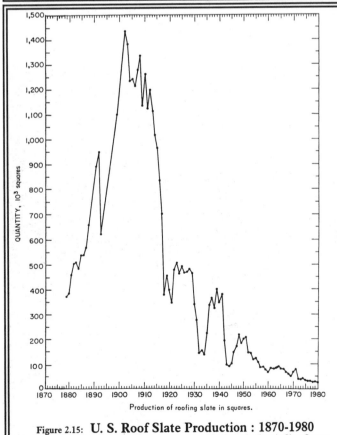

Production of roofing slate in squares.

Figure 2.15: **U. S. Roof Slate Production : 1870-1980**

Source: Moebs, Noel N., and Marshall, Thomas E., 1986; Geotechnology in Slate Quarry Operations, Bureau of Mines Report of Investigations #9009, US Dept. of the Interior, p. 9.

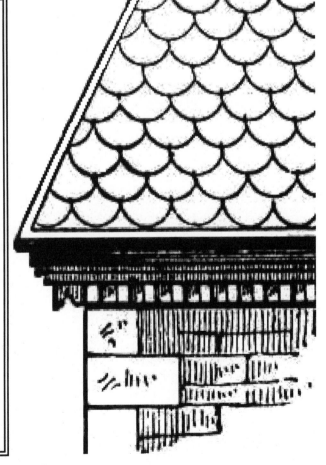

Figure 2.16: ▲ Very old, dilapidated, slate roof in Europe. Each slate was hung on a thin lath (visible above) with a single wooden peg as thick as a pencil and about two inches long.
▼ Same sort of roof, relatively intact, on a pigsty in Wales.
(Photos by Dave Starkie)

Figure 2.17

▲ These old European monastery buildings have roofs made of limestone slabs, not slate.
▼ A closeup of one of the limestone slab roofs.

(Photos by Dave Starkie)

Figure 2.18

▲ Not all stone roofs are slate roofs! The above house in Herefordshire, England, is roofed with local stone, not slate. ▼ The house below, located in Yorkshire, England, is roofed with Yorkshire stone, which is similar to sandstone. In the USA, however, almost all stone roofs are slate roofs.

(Photos by Dave Starkie)

Figure 2.19: *Not* a slate roof, this Cotswald stone roof typical of Gloucestershire, England is split like slate, but the roof shingles are generally smaller and thicker. Roofs like this are typically pegged, i.e. the stone is not nailed in place, but instead is hung on lath by wooden pegs driven (like nails) through holes in the stone.

(Photo by Dave Starkie)

Figure 2.20 Slate roof in Galicia, Spain
(Photo by Dave Starkie)

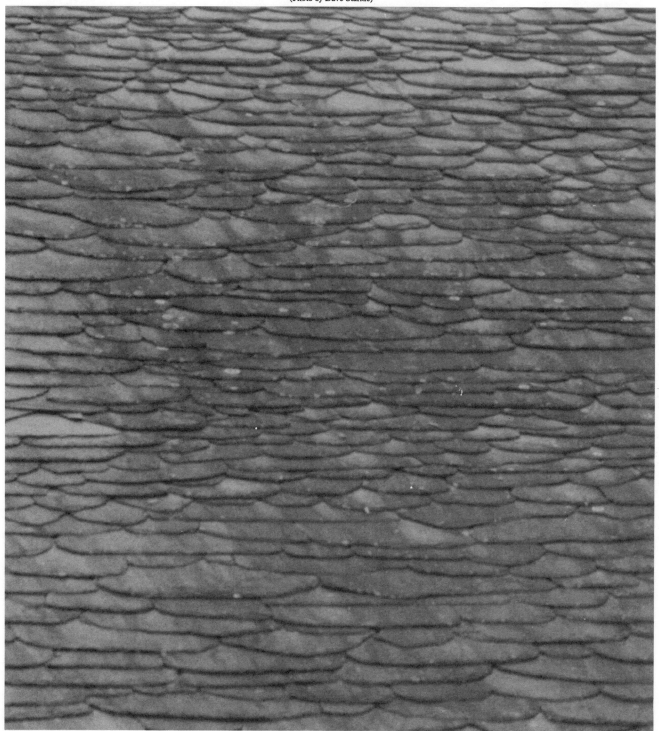

References -- Chapter 2

• Behre, Charles E., (1933), <u>Slate in Pennsylvania</u>, PA Geological Survey, Fourth Series, Bulletin M 16. p. 29.

• Bowles, (1934), <u>The Stone Industries</u>, First Edition, McGraw-Hill Book Co., Inc., New York and London, p. 230.

• Levine, Jeffrey S., (1993), *"The Repair, Replacement and Maintenance of Historic Slate Roofs,"* U.S. Department of the Interior, National Park Service, Preservation Briefs # 29.

• Moebs, Noel N., and Marshall, Thomas E., 1986, Geotechnology in Slate Quarry Operations, Bureau of Mines.

• Pennsylvania State University, School of Mineral Industries, <u>Properties and New Uses of Pennsylvania Slate</u>, The Pennsylvania State College Bulletin, Volume XLI, No. 30, July 18, 1947, pp 7-25, 134-137.

A Cotswald stone roof (not slate). Note the rounded stone valley, a common European style. An exposed metal valley is typical of most American slate roofs.

(Photo by Dave Starkie)

Chapter Three
Identifying Roof Slate
(Understanding What's Overhead)

In order to understand an antique slate roof and its historical significance, as well as its expected longevity and overall value, one must first identify the type of slate on the roof. That simply means one must determine where the slate was quarried. Slate varies significantly from quarry to quarry and slate region to slate region. As we've seen, some quarries produce soft slate, and some produce hard slate. In general, slate can be expected to last from 55 years to centuries on a roof in the United States, depending on the *type* of slate. Some slates will last significantly longer than 400 years if split thicker than the standard 3/16". However, our concern, first and foremost, is to determine where the slate came from that is now on the roof.

A good slate roofer can tell what type of slate you have at a glance, but you'll have to figure this out on your own if an experienced slater is not readily available. Don't be disappointed if you can't find anyone to tell you what kind of slate you have. Remember that the men who installed the old slate roofs are almost all dead and gone, and most younger roofers today can't be

Figure 3.1:

▲ The date on this Vermont "sea green" slate roof was made by using red slate found in the Granville, NY, area. This roof was 113 years old when photographed and, if properly cared for, could still last for generations.
(Photo by the author)

Figure 3.2: ▲ **Slate roof on barn - Lake District, northwestern England. Some of these massive slates are a full inch thick.** (Photo by Dave Starkie)

bothered with trying to find out where roof slate came from, especially as most of them don't work on slate roofs anyway.

The easiest way to identify roof slate is by visual identification. Look at the color and smoothness. Next, handle the slate, break it apart and judge its strength and density. Old soft slate may almost fall apart in your hand, and should be visibly flaking. Hard slate won't, and isn't. Then, after breaking open a piece of slate, look at its interior. The interior of the slate is the same color as when it was originally quarried. The exterior of some slate changes color over time with exposure to the weather, making it hard to identify with certainty without breaking a piece open. Some slate roofs have been completely tarred over by well-meaning but misguided roofers, and the only way to identify these slates is by breaking one open (use a pair of pliers and break a small piece off an edge of the roof). It may also be helpful to remove a slate from the roof and look at the back side.

In any case, you can look at a slate till you're blue in the face, but if you don't know what to look for it won't do you a bit of good. Figure 3.3 shows a map of the eastern seaboard of the United States, indicating the five main slate producing regions. There's nearly a 100% chance that the roof came from one of these regions. Otherwise, the slates may have been imported from Europe. On older roofs this is not likely except on very expensive establishments. Besides, some

In general, slate can be expected to last from 55 to 400 years on a roof in the United States, depending on the type of slate, and proper maintenance. Slate type depends on where the slate was originally quarried.

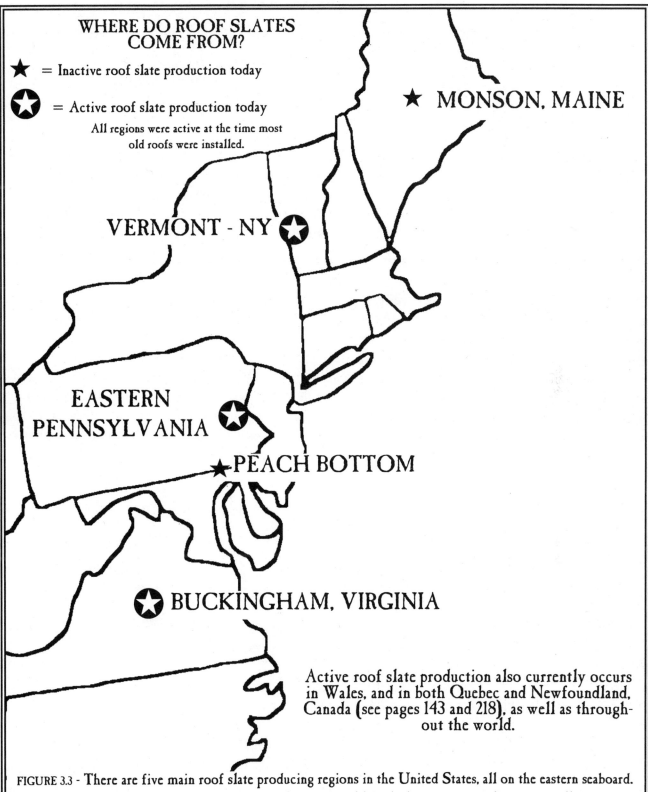

WHERE DO ROOF SLATES COME FROM?

★ = Inactive roof slate production today

✪ = Active roof slate production today

All regions were active at the time most old roofs were installed.

★ MONSON, MAINE

VERMONT - NY ✪

EASTERN PENNSYLVANIA ✪

★ PEACH BOTTOM

✪ BUCKINGHAM, VIRGINIA

Active roof slate production also currently occurs in Wales, and in both Quebec and Newfoundland, Canada (see pages 143 and 218), as well as throughout the world.

FIGURE 3.3 - There are five main roof slate producing regions in the United States, all on the eastern seaboard. Only three of these regions are producing roof slate today, although there are many slate roofs still in existence bearing slates from all regions. The regions that are no longer producing roof slates have not quit because they've run out of slate. Quite the contrary, the quantity of slate that has been removed from the earth has so far, in all regions, only scratched the surface of the immense deposits. Roof slate production has stopped in the Monson, Maine region and the Peach Bottom, Pennsylvania region due to economic factors: the roof slates have simply become too expensive to produce.

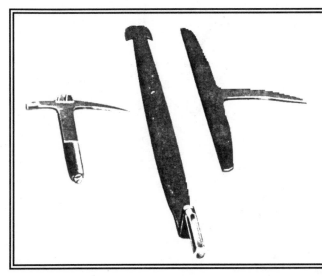

Figure 3.4 SLATER'S TOOLS

Slate Hammer *(left)*
For cutting, hammering and punching holes in roof slate.

Slate Ripper *(center)*
For removing slates from a roof

Slater's Stake *(right)*
To aid in the cutting of a roof slate with a slate hammer.

of the best roofing slate in the world comes from the United States, so even on the more expensive homes the slate probably came from the U.S. There are a few other areas of the United States that have produced quantities of roof slate, and if you live near one of them (see Figure 3.5) you may have that type of slate on your roof.

For those who live in any of the main slate producing regions, your identification problems are more than likely solved. If you live near Monson, Maine, you can be pretty certain you have Monson slate on your roof. Same goes for Peach Bottom, Vermont-New York, Buckingham, and Eastern Pennsylvania. In visiting all of these slate quarrying regions I wasn't surprised to see that in each region they used their own slate on their own roofs. If you don't live in any of these regions and have a slate roof, then you'll have to familiarize yourself with the types of slate each region produces.

Let's start at the beginning: slate has several basic colors, namely black, "blue-black," gray, purple, green, "sea green," and red. You know what **black** looks like (like a blackboard), although some black slate turns brownish with age, and some develops white overtones; **blue-black** looks like black with a slight hint of blue; **green** slate does have a greenish tint to it when quarried, but turns light gray with age; **sea green** looks blue-green when quarried, but turns light gray with mottled hues of red, tan, pink, or orange (depending on your eyesight) with age; **purple** slate is dark purple when quarried and stays that way with age, but may look black or dark gray to an untrained eye (some purple slates have green streaks in them, **variegated purple** has lots of green streaks); **red** slate is red like terra cotta ceramic tile, and stays red with age; **gray** slate is somewhat light gray when quarried and may have black streaks, and it retains its original color with age. The industry term for slate that changes color with age and weather is "fading" or "weathering" slate, and slate that does not change color is called "unfading."

Each region produces colors of slate peculiar to that region, as follows:

MONSON, MAINE: Solid black slate with slight luster, hard, durable. One of the best slates available. May last hundreds of years. These slates are no longer quarried, so recycling old roofs is imperative.

VERMONT/NEW YORK BORDER: Green, sea green, gray, red, purple. Many of the *colored* slates come from this area. Most of these slates are of exceptional quality and very durable and can be expected to last between 150 to 400 years. These slates are still being quarried and can readily be bought commercially.

EASTERN PENNSYLVANIA: Black, blue-black, dark gray, and various other shades of black. Virtually all of the soft slates come from this region, with lifespans of as little as 50

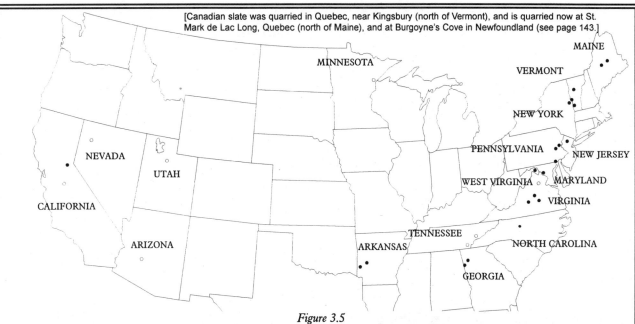

[Canadian slate was quarried in Quebec, near Kingsbury (north of Vermont), and is quarried now at St. Mark de Lac Long, Quebec (north of Maine), and at Burgoyne's Cove in Newfoundland (see page 143.)]

Figure 3.5

LOCATION OF KNOWN SLATE DEPOSITS IN THE UNITED STATES

Solid circles are (or were) productive districts, hollow circles are prospective districts. The five commercially important U.S. roof slate districts are:
1) Monson District, Maine; 2) New York/Vermont border; 3) eastern Pennsylvania; 4) Peach Bottom (PA/MD border); 5) Buckingham District, Virginia.

<u>MAINE</u>: Monson, Brownsville, North Blanchard. Solid black slate with slight luster, (hard, durable).

<u>VERMONT/NEW YORK BORDER</u>: Rutland County, Vt. (Poultney, Fair Haven, Wells, East Poultney) and Washington County, NY (Granville). Green, sea green, gray, red, and purple slates, (hard, durable).

<u>EASTERN PENNSYLVANIA</u>: Lehigh and Northampton Counties (Bangor, Pen Argyl, Chapman, Slatington, etc.) Black, blue-black, dark gray, other shades of black, (ranging from soft to moderately hard).

<u>PEACH BOTTOM, PA/MD</u>: York and Lancaster Counties, Pa. and Harford County, Md. Black slate, (very hard and long lasting).

<u>BUCKINGHAM, VA</u>: Buckingham County (Arvonia). Black slate, (very hard and durable).

<u>NEW JERSEY</u>: Sussex Co. (Lafayette). Black, same slate as easternmost Pa.

<u>GEORGIA</u>: Polk County (Rockmart), and Bartow County (near Fairmount). Greenish gray hard slate south of Fairmount; bluish gray (black) slate at Rockmart, ranges from hard to somewhat soft.

<u>TENNESSEE</u>: Monroe County (Tellico Plains). Purplish, greenish and black slate (hard).

<u>MINNESOTA</u>: Baraga County (Arvon). Black slate (quality uncertain).

<u>UTAH</u>: Slate Canyon. Green and purple slates (quality uncertain).

<u>ARKANSAS</u>: Montgomery County near Norman and Slatington. Red and green slate, and some greenish gray and black slate near Mena, Polk County (quality uncertain).

<u>CALIFORNIA</u>: Eldorado County, near Kelsey. Dark gray slate resembling Pa. slate in color (quality uncertain).

<u>NORTH CAROLINA</u>: Quality and type uncertain.

Note: The original reference materials included no descriptive information for the States shown on the map that were not listed above.

[From a 1914 map by T. Nelson Dale for the U. S. Geological Survey, Bulletin 586, Plate 1; North Carolina data from Moebs, Noel N., and Marshall, Thomas E.,; 1986; Geotechnology in Slate Quarry Operations, USDI, Bureau of Mines RI 9009]

> *Slate has several basic colors, namely black, gray, purple, green, "sea green," and red. The industry term for slate that changes color with age and weather is "fading" or "weathering" slate. Slate that does not change color is called "unfading."*

years or less, although some of the black and gray slate from this area can be quite long-lasting (100-150 years). There is a considerable variety of blackish slate from this region, including **CHAPMAN** slate (a striated, durable black slate), and "Cathedral Gray," a lighter, durable black. "Ribbon" slate also come from here, having a carbon band across their face that can cause the slate to break with age.

PEACH BOTTOM: Black slate, very hard and long lasting. This is an excellent slate with a lifespan that may approach 400 years, and should be preserved on roofs, or recycled, as this region is no longer quarrying roof slate. On close examination you may see a sparkly luster to the slate, especially on the back (unweathered) side.

BUCKINGHAM: Black slate, very hard and durable, will last centuries. Similar to Peach Bottom, will sparkle in sunlight on close examination, especially on back where air pollution hasn't stained the surface.

Let's throw a monkey wrench into this situation now that you've learned which slate comes from where. *Air pollution* causes slate to change color with age. Lots of slates are hard to identify simply because of the neighborhood the roofs are in. Smoke from coal stoves or from local factories will cause the roof to develop a peculiar hue which may turn a black slate lighter gray, or a green slate brown. There's no way to predict this, so in these situations you'll have to

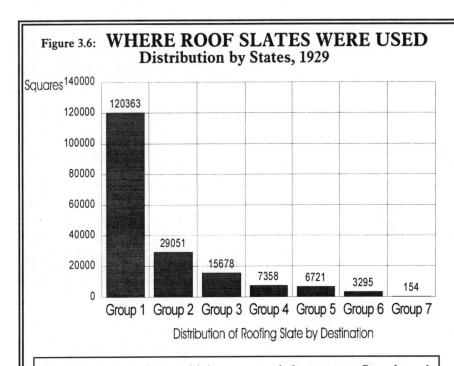

Figure 3.6: **WHERE ROOF SLATES WERE USED**
Distribution by States, 1929

Squares

Group 1: 120363
Group 2: 29051
Group 3: 15678
Group 4: 7358
Group 5: 6721
Group 6: 3295
Group 7: 154

Distribution of Roofing Slate by Destination

The above graph shows which states used the most roofing slates in 1929 (group 1), and which used the least (group 7). This provides a rough estimate of where the majority of slate roofs are located today. Note that the southern states seem to have the least number of slate roofs. This may be due to the heat of the south, which can be absorbed by the stone roofs, or to a lack of proximity to prolific quarries.

[Source: Bowles, (1930), *Slate in 1929*]

GROUP 1
Delaware, District of Columbia, Maryland, New Jersey, New York, Pennsylvania, West Virginia

GROUP 2:
Connecticut, Maine, Massachusetts, New Hampshire, Rhode Island, Vermont

GROUP 3:
Illinois, Indiana, Michigan, Ohio

GROUP 4:
Alabama, Kentucky, Louisiana, Mississippi, Tennessee

GROUP 5:
Arkansas, Iowa, Kansas, Minnesota, Missouri, Nebraska, Oklahoma, Texas, Wisconsin

GROUP 6:
Arizona, California, Colorado, Idaho, Montana, Nevada, New Mexico, North Dakota, Oregon, South Dakota, Utah, Washington, Wyoming

GROUP 7:
Florida, Georgia, North Carolina, South Carolina, Virginia

Figure 3.7: These new Vermont roof slates are "special order" slates split extra thick. Standard roofing slate thickness is much thinner at 3/16 inch.

(Photo by the author)

Figure 3.8:
A new slate roof with unique, overlapping, interlocking slate ridge in South Gallicia, Spain.

(Photo by Dave Starkie)

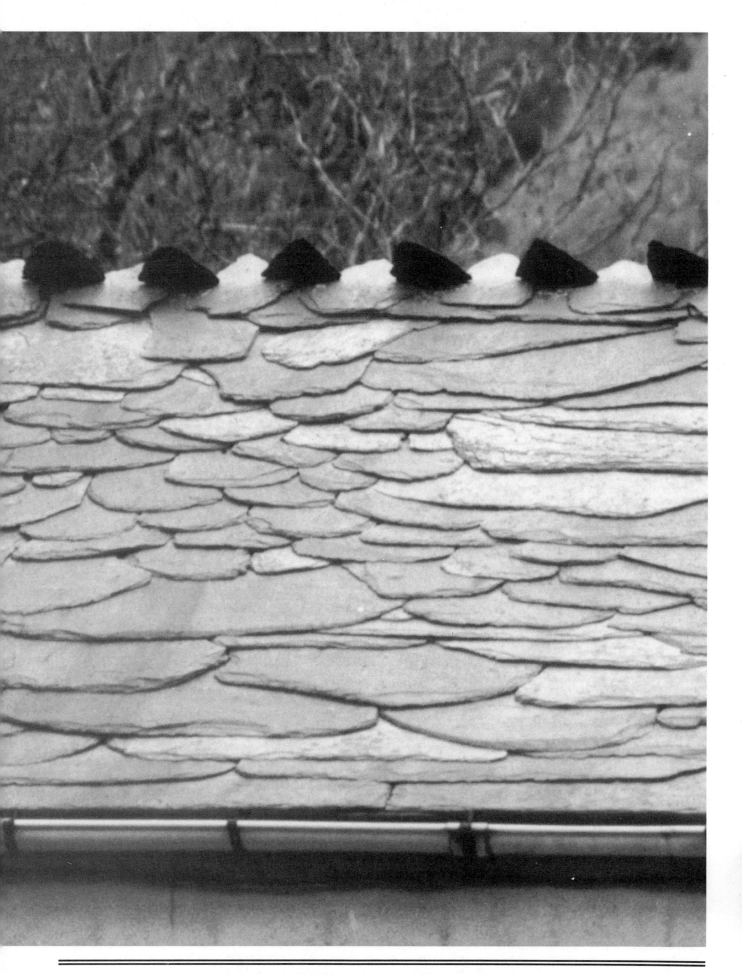

resort to breaking open a slate and looking inside to identify the color. Remember, the color inside the slate indicates the color at the time of quarrying. Add to this the fact that some slate turns color with age, and then may undergo another color change due to air pollution, and you can see why visual identification can be tricky at times. Furthermore, some black slates from eastern PA will exude a chalk that turns them somewhat white (usually around the edges), while other black slates from the same region will weather to a brownish hue.

To add to the confusion, trees change the color of a slate roof. Leaf drop from an overhanging tree will stain slate, making them a darker color. Also, slate is readily stained to a very dark brown by rusting metal, such as rusting ridge iron or flashing.

Why is it so important that we identify the slate on old roofs? Well, if the roof has Monson, Vermont/New York, Peach Bottom or Buckingham slate on it, then it's highly restorable and the slate could very likely last the life of the building it's on-- such roofs should practically never have to be replaced. On the other hand, if the roof is made of the softer, eastern Pennsylvania slate, it may have reached the end of its life (after a century) and no amount of work, money or prayer will save it. The roof will have to be replaced whether you like it or not

Figure 3.9: Delabole slate roof in Cornwall, England
(Photo by Dave Starkie)

(although many older soft slate roofs still have *decades* of life left in them). When soft slate roofs are replaced, by the way, they should be replaced with *slate* roofs (not asphalt shingles), and eastern PA black slate is an excellent choice if someone wants a roof that will last a century.

In any case, it's important to know how to identify the slate on a roof. In subsequent chapters we'll look at each of the slate regions in greater detail, so if you're confused about the different types of slate, there's still hope!

But first, we should look at how and why slate was discovered and quarried in the United States, and in order to do that we must take a trip to Wales.

QUICK REFERENCE - IDENTIFYING OLD AMERICAN ROOF SLATES

ORIGIN	COLOR**	APPEARANCE WHEN OLD	LONGEVITY Manuf. Estimate	Actual (Estim.)	AVAILABLE NEW TODAY?	SUBSTITUTE	RECYCLABLE WHEN OLD?
MAINE (page 139)	Black	Smooth, sheen, no flaking.	?	200 years +	no ··········	1.Lehigh-Nor. 2.Spanish Black 3.Buckingham	yes
VERMONT (page 77)	Green	Light gray, smooth	75	200-250	YES★	4. Glendyne (Canada)	yes
	Purple	Dark purple, smooth	75	200-250	YES★		yes
	Black	Mottled black, "	75	?	YES★		yes
	Sea Green	Light gray+rust some flaky	75	150-200	YES★		yes
NEW YORK (page 77)	Red	Brick red, smooth, no flaking.	75	200-300+	YES★		yes
PENNSYLVANIA (page 99)							
Leh/North. (page 99)		Various black shades, flaky, crumbly. May be brownish on surface, or have whitish edges.	50	50-125	YES★		no
Peach Bottom (page 115)		Black, smooth, sheen (may be only visible on back of slate), hard, no flaking.	150	300+	no	Buckingham	yes
VIRGINIA (page 127)	Black	Gray-black, smooth sheen, hard, no flaking.	150	300	YES★		yes
GEORGIA (page 132)	Black	Gray to black, maybe flaky	?	75-200	no	See 1, 2, 3, 4 above	

★FOR SOURCES OF NEW SLATE SEE PAGE 218
★★FOR COLOR IDENTIFICATION SEE CENTER COLOR PHOTO SECTION

References - Chapter Three
• Bowles, Oliver, and Coons, A. T., 1930, Slate in 1929, U. S. Department of Commerce, Bureau of Mines, Mineral Resources of the United States, 1929 - Part II (Pages 161-74).

50   Entrance to the now abandoned Vivian Slate Quarry near Llanberis, Wales.

Chapter Four
Wales

"In 1980, a small boat, loaded to the gunwales with slate, was discovered at the bottom of the Menai Straits. It has been dated at 1270, and, although no one can be certain, the slates appear to have come from the Penrhyn Quarries at Bethesda, North Wales."

John Brigden

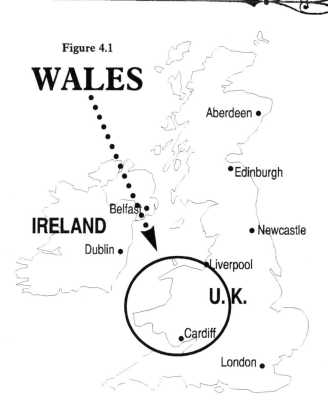

Figure 4.1

WALES

IRELAND

U.K.

A study of slate quarries and the phenomenon of the slate industry in the United States starts in Wales. In order to gain a deeper and more thorough understanding of slate roofing, we should take a look at the men and women, the times and the places that started it all. In doing so, we shall understand that slate quarrying in America, and the use of slate for roofing, were not new endeavors brought to fruition by pioneering, resourceful people in a new country. They were instead ancient skills wrought from the knowledge of a timeless people; a people who endured lengthy sea voyages to carry their mastery of stone work across the vast Atlantic Ocean to a new land of hope and promise. And with their proud trades they brought proud traditions: the Welsh language, church, and customs. In essence, they brought Wales itself, carving out small enclaves in America that seemed very much like their former home.

The mountainous country of Wales is located along the western edge of England and measures 136 miles in length, varying from 37 to 92 miles in width. North Wales is home to Mt. Snowden, the highest mountain in the UK, a rugged, snow-capped mountain which practically casts its shadow over the largest slate quarries ever worked. These quarries are laden with immense slate deposits yielding some of the best purple, green and black slate in the world.

Pangaea broke apart hundreds of millions of years ago, shifting the earth's crust eons before the earliest evolutionary beginnings of the human race. That separation of the planetary surface left a link between Europe and the United States in the form of underground slate deposits that would eventually draw the Welsh people to the new land of America, and in so doing influence the ethnical and cultural development of a country we now call home.

"Keeping up with the Jones's" is a common American saying, and Jones is a Welsh name. So is Thomas, Roberts, Evans, Hughs, Lewis, Williams, Phillips, Powell, Jenkins, Griffith, Richards, Morgan, Humphries, Edwards, Davies, Hopkins, and others. Granted, slate is not the only reason the Welsh emigrated to the United States. They were also master tin platers, coal miners, and iron workers. However, their expertise in slate dates back centuries.

The most important slate regions in Wales are in the ancient Kingdom of Gwynedd (pro-

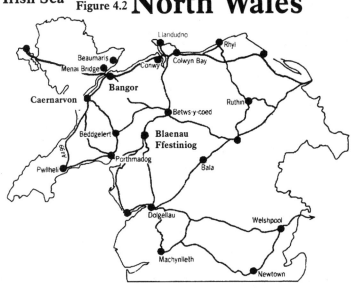

Irish Sea Figure 4.2 **North Wales**

FIGURE 4.3 **CAERNARVON AND NEARBY QUARRY AREAS**

nounced "Gwineth" - the Welsh "dd" is pronounced like the English "th"), including Snowdonia in North Wales, the home of Mt. Snowdon which rises 3,560 feet above the Irish Sea to the northwest. There are also fourteen other peaks rising over 3,000 feet in this area, making it a rugged territory where the removal of tons of stone was not just a challenge, but a feat. This harsh terrain in North Wales is particularly notable when one considers the primitive transportation systems available until recent times, and the poor conditions of the roads (if any roads at all) in the nineteenth century when most of the slate was removed.

Up until the end of the 1700s, sledges without wheels were used to transport goods throughout the area, and in 1798 Reverend Richard Warner described them as having *"the shape similar to the body of a waggon, capable of containing two or three hundred weight of peat."* These sledges were drawn by a Welsh pony, a small beast of burden owned by almost every cottager.

North Wales was also geographically isolated from both England and South Wales by marshland, moorland, estuaries, and the Severn River. Or, as one quarry manager put it, *"[South Wales] is a foreign country as far as the North Walian is concerned."* The uplands of Caernarvonshire were suitable only for grazing sheep, or for the scenery, and visitors have migrated there for centuries in order to get a glimpse of the magnificent countryside which boasts seaside and mountain range, green valleys and quaint villages.

What this all adds up to is a slate industry that, although somewhat active for centuries, did not boom until the 1830's. Once the roads and rails were in place the industry expanded rapidly, only to decline just as suddenly in the early 1900's. The annual output of slates in North Wales in 1832 was 100,000 tons, which rose to about 450,000 tons in 1882. By 1972 the output had fallen to about 22,000 tons due to market competition from ceramic tiles and concrete roofing materials (not from asphalt shingles, by the way, which are a cheap, temporary, petro-chemical roofing material extremely popular in the United States but almost unheard of in Europe).

One early description of the resources of Wales dates from 1387, in the words of John Trevisa as translated from Ranulf Higden:

> *"Valeys bryngeth forth food, And hills metal right good,*
> *Col groweth under lond, And grass above at the hond,*
> *There lyme is copious, And sclattes also for hous."*

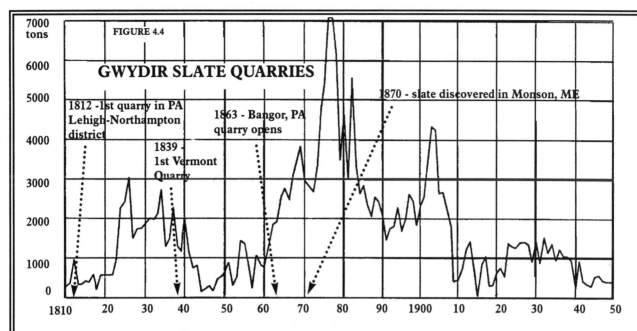

SLATE PRODUCTION IN GWYDIR QUARRIES, WALES, 1810 - 1950

Although this graph only shows the production of one *small* Welsh quarry (Gwydir), it follows a *pattern* much in line with other areas of Wales. Note the slump in production between 1840 and 1860. It is perhaps no coincidence that much of the early immigration of quarry workers from North Wales to the United States took place during the mid-1800's. The Gwydir slate quarry was a small quarry which produced about 7000 tons of slate at its peak. In comparison, Wales' largest quarry, the Penrhyn Quarry, which was responsible for the production of one quarter of all Welsh slate, produced 130,000 tons in 1862. The Ffestiniog Quarries produced 145,000 tons in 1882.

AND IN AMERICA

Slate in America was discovered at Peach Bottom by the Welsh in 1734, and the first commercial quarry opened there in 1785. The first slate quarry in the Lehigh-Northampton district of eastern Pa. opened in 1812. Owen Jones, a Welshman, discovered slate along the Lehigh River in 1846, and slate was subsequently discovered near what is now Bangor, Pa, in 1853. The Bangor Quarry was opened by a Welshman named Robert Jones in 1863. Slate was discovered in Vermont by a Welshman in 1835, and in 1870 slate was discovered near Monson, Maine, once again, by a Welshman.

[Source of graph: Williams, M. C., and Lewis, M. J. T., 1989, <u>Gwydir Slate Quarries</u>, Snowdonia Nat. Park Cr., Plas Tan y Bwlch, Maentwrog, Blaenau Ffestiniog, Gwynedd LL41 3YU, Wales]

The word "slate" comes from the Middle English "slat" or "sclate," and after around 1630 the latter version became exclusively northern or Scottish. The word is related to the French word "esclater" which means to break into pieces, and refers to the cleaving characteristic of the rock. Therefore, in the verse we just read, the last line refers to the abundance of slate used for roofing as well as other house construction purposes (such as walls) in Wales.

In Welsh, cleavable rocks such as slate can be called "llech" (the "ll" is pronounced like a guttural "cl"), and the name is reflected in the name of at least one slate quarry, the "Llechwedd Quarry" near Blaenau Ffestiniog. The older Welsh name specifically referring to slate is "ysglatus," "ysglats," or "sglatys," and dates from the fifteenth century. One poet of the time (late 1400's) referred to Welsh slates as *"jewels from the hillside,"* and *"warm slabs, as a crust on the timber of my house."* The cleavage of the slate is called "hollt" in Welsh.

Slate is located in Wales in the counties of Caernarvon, Denbigh, Merioneth, Montgomery, and Pembroke, although the largest quarries are located in Caernarvonshire (Caernarvon County), which also is home to the slate exporting cities of Caernarvon and Bangor. The two largest slate quarries in the world are located in Caernarvonshire, and are known as the

Figure 4.5 <u>Analysis of Typical Slate of North Wales (%)</u>

Silica 55.30
Oxide of Iron .10.00
Alumina 24.84
Lime 0.36
Magnesia 2.46
Carbonic Acid .nd
Sulphuric Acid 0.21
Potash 1.47
Soda 0.53
Water of Hydration 4.70

[Source: Lindsay, Jean, (1974), <u>A History of the North Wales Slate Industry</u>, p. 292]

Penrhyn Quarry and the Dinorwic Quarry, located on opposite sides of the Elidir Fawr Mountain. The Penrhyn Quarry, which is still in production today, eventually grew into a hole a mile long, a third of a mile wide, and 1,300 feet deep, and out of this hole 300 million tons of slate and rock have been removed, mostly by hand. The Dinorwic quarry was so large it had as much as fifty miles of rail track within it, covering a total area of about 700 acres, and eventually leading to the removal of an entire mountaintop.

The slate in central Caernarvonshire dates from the Cambrian age (600 million years old), making it a very hard, and durable material. The slate towns in this area include Bethesda (the Penrhyn Quarry), Llanberis (the Dinorwic Quarry), and Nantlle (the Dorothea Quarry).

A second slate producing area exists near Blaenau Ffestiniog in Merionethshire and in the adjacent parts of Caernarvonshire (see Figures 4.1, 4.2, 4.3). Here the slate is of Ordovician age, which, at roughly 500 million years old, is younger than Cambrian, although still very durable. A slate worker at the Llechwedd Quarry once told the author that Blaenau Ffestiniog slate would last 400 years. These are the most important Ordovician slates in the Snowdonia region, and the deposits dip deep into the earth under the mountains, so most of the slate has been deep mined. The Llechwedd Quarry dates from 1846, and boasts of a valuable "old vein", which was purported by a worker there to once have produced thin sheets of slate long enough and pliable enough to bend into a circle, although this claim is found unbelievable by American slate quarriers. The underground tunnels in this mine totaled 25 miles in length, and descended to 900 feet. Some of the original tunnels have been restored for historical purposes and can now

be toured on an electrically powered underground tram for a distance of about a half mile (see Figure 4.6).

There are also three other commercially important slate producing areas of Wales: the country between Towyn and Dinas Mawddwy, the country between Llangollen and Corwen, and the Presely district of Pembrokeshire. Incidentally, slate is also located in England in Cornwall, Devon, the Lake District, Aberdeenshire, and Argyll.

Welsh slate can be of a variety of colors in addition to purple, green and black, going by such quaint names as silky red, hard spotted blue, royal blue, curly red, old blue, red hard, hard red bronze, purple green wrinkled, sage, willow, barred red, old quarry blue, and blue grey mottled in the Llanberis area alone. The Nantlle area has a green vein, silky

Figure 4.6: ▲ A sea of slate roofs best describes the town of Blaenau Ffestiniog, which rests stoicly amidst the barren, misty mountains of North Wales.

▼ A tunnel in a Blaenau Ffestiniog slate quarry. Some descend to 900 feet below the surface. Many have rail tracks for slate cars and trams. The tunnels open into cavernous rooms where the slate was removed by hand, in candlelight, then winched to the surface and split into shingles. (Photos by the author)

NAME	SIZE
Kings	20" x 36"
Queens	20" x 34"
Princess	14" x 24"
Duchess	12 "x 24"
Marchioness	11" x 22"
Countess	10" x 20"
Viscountess	9" x 18"
Lady	8" x 16"
Doubles	7" x 12"

Figure 4.7: Individual names were once assigned to each standard size of roof slate. In 1933, however, the British Standards Institution decided to dispense with such names and define sizes instead by length and width.

Source: Brigden, "Turning Stone into Bread"; and Lindsay, Jean, (1974), A History of the North Wales Slate Industry, p. 295; and Ffestiniog Slate Quarry, Blaenau Ffestiniog, Wales (displayed on a wall - see next page)

Discrepancies existed when these names were used. Some sources listed ladies as 10x16, some as 12x14, small ladies as 8x14, wide ladies as 12x16, and similar variations existed with the other names, which is probably why this nomenclature was abandoned.

vein, blue mottled vein, red and blue striped vein, red spotted vein, and red vein. The Penrhyn area has green, purplish blue, grey, grey mottled, mottled and striped, and blue. The Ffestiniog area has mainly blue-grey. In Corris, Abergynolwyn, and Aberllefenni the slate is pale blue, and in Glyn Ceiriog and near Llangollen the slate is blue. Mind you that these descriptions are in industry terms which tend to describe black as "blue," and the slates they describe as red and blue striped slates do not look like the American flag. In fact, the slates from Wales can be generally categorized as dark purple, light green or black, although some of the purple is reddish and some bluish, and variations exist in all the colors. In short, Welsh slate is quite similar to some of the slate of the American continent, especially Vermont slate, and it is identical to the slate of eastern Canada, at Burgoyne's Cove in Newfoundland.

HISTORY

One of the oldest slate quarries in Wales is the Cilgwyn Quarry in Nantlle, dating from the 12th century. Other records mention a slate roof installed in 1317 on the huge Caernarvon Castle by "Henry le Sclatiere." In 1399, a fellow named Creton, in reference to a trip to a Welsh town just north of Bangor, (Conway), wrote:

"So rode the King, without making noise,
That at Conway, where there is much slate on the houses,
He arrived with scarce a pause,
At break of day."

Records show a man named Sion Tudor ordering 3,000 slates in 1580 from Bangor, Wales, to replace the thatch roof on his house. In 1682, parts of the St. Asaph Cathedral were slated with Penrhyn slates, and in 1930, nearly 250 years later, the same slates were still good enough to be replaced on new wood on the same roof when the roof was redone.

A record of a lease in Wales during the period 1568-74 reads:

"I did demise unto Gruffith one tenement called Lloyn y bettws . . .20 shillings to

me towards the slating of the dwelling house, and he to send the carriage of the slates and the meat of the slaters and carpenters."

It's anyone's guess what "meat of the slaters" is supposed to mean (probably their dinner); the point, of course, being that slate roofs have been common in Wales for centuries. Other records speak of slate roofs in 1536-9 when John Leland stated, *"The houses within the town of Oswestre be of tymbre and slated,"* and *"they dig oute slate stones to kyver houses,"* and in 1597, when John Wynn's memoranda included a note to *"slate the cattle houses."* Such records continue through the 1600's.

In the late 1500's slate was being exported to Ireland from Wales, and about 100,000 were exported in 1587. A hundred years later, ten times that amount would be sent over. Other shipping records show slates being exported from North Wales throughout the 1700's, and between 1729 and 1730 over two and a half million slates were sent from Welsh ports, a million of these to Ireland.

The majority of the medieval roofing slates that have been examined have square corners, with sides that are more or less parallel, and are quite small and thick. Their usual size is about seven inches by three and a half to four inches. Today, such tiny slates are not available and even a six inch by twelve inch slate is considered quite small. The early Welsh slates remained small until the skill and the expertise of the slate workers increased. In 1740 the slates doubled in size, and later doubled again. These slates were called "doubles" and "double-doubles." In the 1500's, Welsh shipping records showed slates ranging in size from 5" X 10" to 6" X 12" and 1/2 to 3/4 inch thick. In contrast, most slate roofs installed in America during the late 1800's and early 1900's have uniform slates (all the slates on the roof are the same size) ranging from 9" X 18" to

Figure 4.8: The author at the Blaenau Ffestiniog museum where the various sizes of roof slate made in Wales over the years are displayed. Sizes range from 10"x4" to 36"x20". (Photo by the author)

Figure 4.9:

▲ The Welsh Slate Museum at Llanberis, built and roofed of slate, was once the mill where slate from the huge Dinorwic quarry was processed. Note the mountain of slate rubble in the background. The museum now displays the old tools and equipment, including the 50' diameter water wheel which powered the mill's 1/8 mile of line shaft.

▶ Dafyyd Davies a museum worker, demonstrates the rhythmical motion needed to trim slates with a "stool and traverse." The "slate knife" he holds can still be bought in hardware stores in Wales today.

(Photos by the author)

14" X 24", and averaging 3/16" thick.

Slate had other uses in Wales besides roofing. Many houses and buildings had walls built of slate stones, and still stand today in perfect condition. A slate spindlewhorl used for the hand spinning of wool was found in Caernarvonshire in 1944 and is now displayed at the National Museum of Wales at Llanberis. Slates were also used as bake stones, probably from medieval times. They were similar to our modern pizza pans, about an inch thick, and were used to bake bread known as "bara llech" or "slate bread."

Ironically, the predominant type of roofing in Wales was not slate until the nineteenth century. Prior to that time it was thatch, which is a thick and beautiful covering of reed that can absorb enough water to become quite heavy, heavier than slate. It's also flammable when dry, and it contributed to some of the great fires of the time, such as the Great Fire of London. It doesn't take too many Great Fires before people start looking for a fireproof roofing material such as slate. At one point in time it was declared in London that, *"Every person who should build a house should take care that he did not cover it with reeds, rushes, stubble or straw, but only with tiles, shingles, board or lead."* Wooden shingles were also used for roofing, and around 1810, Thomas Pennant, while touring Wales, reported that in many regions *"shingles, heart of oak split and cut into form of slates,"* was *"the ancient covering of the country."*

The sale of roofing slate rises proportionally as the construction of new buildings occurs, and when new construction grinds to a halt, as during war-time or economic depression, the production of slate is adversely affected. Slate production in Wales boomed in the 1800's, and the number of slate workers nearly doubled from 1861 to 1881. But in the 1880's a depression in the building trades in Great Britain reduced the demand for roof slate. Although the building industry stagnated in the 1880's, it picked up again in the 1890's, when new houses averaged about 130,000 a year. However, by 1909 new houses had dropped to 90,000 a year, and by 1913 they were down to 62,000. The effect on the slate industry of these fluctuations can be seen on the graph in Figure 4.4.

While the slate industry was subjected to the vagaries of the building trends, it was also affected by foreign developments. Exports of Welsh slate dropped off between 1889 and 1918 from nearly 80,000 tons to 1,500 tons. This was largely due to World War I and the loss of Germany as a customer, as Germany accounted for 72% of Welsh slate exports in 1876, and many buildings in Germany were covered with such slate. The Germans were partial to the black slate of the Ffestiniog area, so their loss as a customer hurt this area the most.

American slates began cutting into the Welsh slate market by the end of the 1800's. In 1898, American slates were being sold in Dublin, Ireland at 25% less than Welsh slate, doing serious injury to the Welsh slate market there. In 1897, Pennsylvania was the leading slate exporter, followed, in order, by Vermont, Maine, Virginia, New York, Maryland, and Georgia. Slates were also being imported into Great Britain from France, Belgium, Norway, Portugal, Italy, the Netherlands, and Germany. The Welsh slate industry fought back against the import of foreign slates by pointing out their disadvantages. They claimed, with considerable exaggeration, that France's blue slates soon turned "dirty grey," while Germany's red slates were "soft as clay," and America's slates were so rough as to be unsuitable for roofing houses. Admittedly, Welsh slate was quite durable and lasted many years even in smoky, acidic city environments, while some American slates "only lasted twenty years," *according to Welsh slate workers*, and some German slates were "even worse." The best French slates were an exception, being much more durable.

In 1901 a depression began that was in part attributed to the Boer War, leading to a scarcity of money and increased unemployment. In 1906, the arrival of asbestos tile on the mar-

Figure 4.10: A "drumhouse" (right) housed a large drum (below), with heavy cables wound against each other. Slate cars attached to the cables were let down from the mountain top on iron rails. The weight of the cars loaded with slate pulled the empty cars back up the mountain. A brake lever controlled the operation. These long abandoned relics remain intact at the Dinorwic Quarry, adjacent to the Welsh Slate Museum. (Photos by the author)

Figure 4.11:
(Bottom left) Looking up the first incline from the bottom of the mountain at Dinorwic. Old rails are clearly visible. The lowest drumhouse, pictured on the previous page, sits at the top of the photo. (Top left) Looking down toward the lowest drumhouse, which now sits around the corner to the right. (Top right) Looking up another incline toward the top of the mountain. Abandoned slate mill sits at right of photo. This quarry (Dinorwic), covering 700 acres, had 50 miles of iron track. Slate walls, some with slate shelters built into them (below), line most pathways.
(Photos by author)

The card of prices (left) is an authentic document from the North Wales slate industry, as are the documents on the following pages related to the attempts to unionize the quarries in the 1800's.

ket further encouraged the downward trend in slate production, which ultimately carried on into the first world war, never to fully recover. An article in a 1910 *Beautiful Homes* magazine referred to this new product as asbestos shingles, made of asbestos fiber and portland cement, and coming in shades of either gray or bright red, while costing no more than a slate roof. Incidentally, many asbestos roofs were also installed in the United States during the early 1900's, and many of them remain on American roofs today. People tend to mistake them for slate roofs, and some even insist on calling them "asbestos slates." However, asbestos roofs are not slate roofs and only vaguely resemble slate - although asbestos roofs are usually good for about 75 years.

The number of Welsh slate mines dropped from 29 to 12 between 1914 and 1918. Building came almost to a standstill during the war, and in 1917, as in the United States, slate quarrying was officially declared a "non-essential industry." As a result, skilled quarry workers were drafted into the military and many quarries closed for the duration of the war.

Although slate is still quarried in Wales today, in many areas, as in the United States, only abandoned quarries remain.

LIVING AND WORKING CONDITIONS

Much of the slate mining in Wales was done in three ways: the slate could be worked in terraces or shelves cut into the sides of a mountain as at Penrhyn or Dinorwic, or large holes or pits could be sunk into the ground as at Dorothea, or the slate could be brought out of the earth in tunnels (Figure 4.12).

When the slate beds plunged deep underground, they had to be followed down through tunnels by the quarrymen in order for the slate to be extracted. The Llechwedd quarry near Blaenau Ffestiniog was one of the deep mines, and the interior of the mine was damp and *totally* dark. Its 25 miles of underground tunnels (Figure 4.6) connected 60' chambers, from which the slate was removed. The workers adhered candles to the rock with clay in order to see what they were doing. The candles were expensive, and often only one candle was shared by two people, especially if the second person was a young apprentice.

The men drove holes into the rock face using six foot long iron bars called "Jwmpars" or

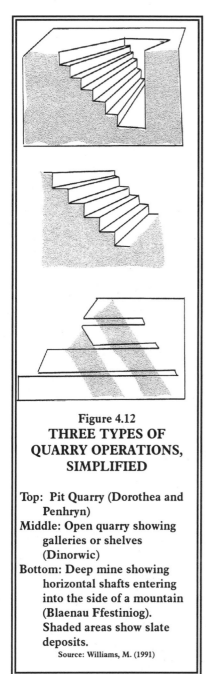

Figure 4.12
**THREE TYPES OF
QUARRY OPERATIONS,
SIMPLIFIED**

Top: Pit Quarry (Dorothea and
  Penhryn)
Middle: Open quarry showing
  galleries or shelves
  (Dinorwic)
Bottom: Deep mine showing
  horizontal shafts entering
  into the side of a mountain
  (Blaenau Ffestiniog).
  Shaded areas show slate
  deposits.
  Source: Williams, M. (1991)

"jumpers" in English. The bar was repeatedly pounded against the rock to produce a hole, progressing six to twelve inches per hour. Blasting powder was tamped into the holes, but not with an iron bar which could spark a disastrous explosion. A fuse would be laid, which burned at a rate of an inch a minute, giving the men time to get out of the cavern and find shelter in the tunnel during the explosion.

Large blocks two to three feet thick and up to ten feet long were brought down by the blast, and were broken up by the rockmen by hammer and chisel, then loaded onto trolleys by a chain hoist and tripod, and winched to the surface. "Badrockmen" were paid by the ton to remove non-productive rock, and "rubblers" cleared the waste.

At Llechwedd, the men worked each chamber in teams of four, two in the chamber and two on the surface in the mill. The two rockmen below the surface extracted the slate and sent it up to their partners above, who broke the slab down further, finally splitting and trimming out the roof slate using only hand tools. Other quarries used four, six, or eight men to a team.

In 1876, a quarry worker wrote, *"When I was a child, I had to walk five miles before six in the morning, and the same distance home after six in the evening; to work from six to six; to dine on cold coffee, or a cup of buttermilk and a slice of bread and butter. Some of the men had to support a family of perhaps five, eight or ten children on wages averaging 60 pence to 80 pence a week."*

Apprentices received no pay for the first six months, and had to wait nine years before they received full pay. Some quarries had their own wood shops, pattern shops, and foundries where they made their own tools, and parts for their equipment.

Men living too far from the quarry to travel home stayed in barracks (Figure 4.13) and traveled home on weekends. The men worked outside during all types of weather, and some quarries provided eating houses, and places with a fire where the men could dry their clothes. Typical meals consisted of tea, egg, and bread and butter for breakfast; tea and bread and butter for lunch; bread and butter for teatime; and potatoes, sometimes with beef, bacon, or buttermilk for supper.

Quarry work was hard and dangerous; over four hundred men and boys have been killed in the Penrhyn Quarry since 1782, and one quarryman, Robert Williams, in 1792 had as his gravestone the piece of rock that fell on him. Fatal accidents at the Bryneglwys Quarry also occurred year-round, and varied in cause, although records reveal typical causes of death: a worker named Owen fell 18 yards down a shaft and died. A few years later, William Owen was knocked over a ledge by a falling rock while boring a hole. He injured his spine and died the next day. A few months later, a piece of rock weighing 7 tons fell on David Evans and Hugh Jones and crushed both of them to death. Ten days later, Edward Davies fell 20 yards while trying to unhook chains from a wagon, and died. This was about the same time another man was killed at the quarry by a runaway truck. Later, John Watkin was killed when the earth above him

Figure 4.13: ▲ Abandoned buildings at the Dinorwic Quarry. Note entire mountainside removed in background.
▼ Workers barracks at Dinorwic, long abandoned - no trace of the roofs remain. (Photos by the author)

Figure 4.14:
Three jobs involved in producing roof slates: the fellow on the right is splitting the blocks into smaller pieces while the man in the center splits them down further into roof shingles. The man on the left is trimming the slate with a slate knife and a "stool and traverse." The Welsh don't punch ("hole") their slates at the quarry as Americans do. They let the roofers do it on the job site.

(Courtesy of Old Line Museum, Delta, PA)

collapsed as he excavated a reservoir, and a few months later Tom Rogers got his coat caught in a rotating turbine shaft and was killed. John Lewis was later trapped by a falling rock and killed; David Roberts was hit and killed by a falling stone weighing 30 pounds; Richard Davies was struck and killed by a rock falling from a chamber roof; David Evans was killed by a rock weighing 25 tons; Owen Ellis fell about 50 feet and was killed when climbing down a chain being used to lower a block of slate; Tom Ellis was killed a few months later when hundreds of tons of rock from a chamber roof fell on him; David Owen fell down a shaft he was exploring out of curiosity, and died; David Davies fell backwards over a ledge when his crowbar slipped as he was working a stone; Edward Lewis died when the ledge he stood upon gave way, crushing him; and numerous others died from similar causes.

In the deep mines, slate had to be left in place as supporting walls in order to prevent undermining and consequent disaster. However, some profit-hungry quarry operators robbed these support walls, thereby thinning them or removing them altogether. In 1883 no less than six

To the late Employés at the Penrhyn Quarries.

I am again instructed by Lord Penrhyn to make you the accompanying offer of work on the same condition as before, viz.:

That there shall in future be no attempt on the part of any Committee to interfere with the management of the Quarry, or to prevent Employés from obeying the orders of the Managers.

Upon the above distinct understanding, all applications (including those of the 71 men who were suspended on 28th Sept., 1896) will be impartially considered, and as many of the late employés as there can be found room for, will be re-engaged, without reference to the events connected with the strike.

As misapprehension appears to exist with regard to the 71 men to whom notice of suspension was given (Sept. 28), I must point out that the suspension of those men (*ipso facto*) ceased immediately the present strike was declared, whereby the men severed their connection with their employer, thus making it obviously necessary for you all to apply individually for re-engagement if you are desirous of obtaining work at the Penrhyn Quarry.

E. A. YOUNG.

NOTICE.

WANTED AT

THE PENRHYN QUARRIES

Blacksmiths, Badrockmen, Boys, Brakesmen, Engine-drivers, Fitters, Foundry-men, Joiners, Journeymen, Loaders, Labourers, Masons, Machiners, Miners, Quarrymen, Rybelwyr, Sawyers, Stokers, Tippers, &c.

Applicants for work can apply on Monday or Tuesday next (February 15th and 16th), between the hours of 10 a.m. and 3 p.m., at the following Offices:

To apply at the Yard Office - - -	Blacksmiths, Fitters, Foundrymen, Joiners, &c.
To apply at the Slab Mill Office - -	Sawyers, &c.
To apply at Tross-y-fordd Office. - -	Badrockmen and Journeymen.
To apply at the Pay Office - - -	Quarrymen and Journeymen.
To apply at the Pay Office - - -	Boys.
To apply at the Marker's Office - -	Loaders, Masons, Miners, Machiners, Engine-drivers, Stokers, Brakesmen, Tippers.
To apply to William Parry, Overlooker's Office - - - - - - - -	Rybelwyr.

Port Penrhyn, Bangor,
8th February, 1897.

E. A. YOUNG.

and a quarter millions tons of overburden and slate collapsed on the workings of the Welsh Slate Company due to irresponsible undermining. The owner of the land, William Edward Oakely, sued the mining company for compensation for the damage done to his assets, and won his claim in one of the longest arbitration disputes in legal history.

Men of North Wales who didn't work in the quarries lived to be an average age of 67 years, while quarry workers were likely to die by the age of 38. Many of the quarrymen suffered from respiratory diseases because of the dust in the quarries. Ironically, quarry medical personnel disagreed. At the Penrhyn Quarry Hospital in Bethesda, medical officer J. Bradley Hughes issued a written statement on June 1, 1922 stating, *"We have no case of Silicosis in this quarry of which I am aware, and I became convinced after four years' experience here that slate dust is not merely harmless, but beneficial."* Dr. E. Shelton Roberts, who had a very long acquaintance with the conditions both at Penrhyn quarry and at other quarries in the district, further maintained that he had not met a single case of Silicosis or lung trouble among the quarrymen attributable to their work at the slate quarries. Dr. John Roberts, general practitioner with 39 years experience in a quarry district wrote, *"During my experience it is remarkable to say that no quarryman has ever complained of cough or choking sensation due to inhalation of slate dust. . . I have not examined slate dust microscopically, but am of the opinion that the particles are soft and none irritant, also it contains very little silicate and a good deal of iron and sulphur, the latter is regarded as a germicide. . . The facts already stated and my experience convinces me, that slate dust is neither an irritant exciting or a predisposing cause of Tuberculosis. There is no doubt that the quarryman's mode of living and poor feeding consisting mainly of tea, bread and butter reduces their resisting power against Tuberculosis."*

Of course, the doctors who issued these statements were on the payroll of the slate quarries, and their reports were used by the North Wales Quarry Proprietor's Association to downplay any detrimental effect on the health of quarry workers caused by the quarry environment. The North Wales Quarrymen's Union responded to the above statements by issuing a statement of their own declaring, *"That we cannot agree with the statement of the case as submitted by the North Wales Quarry Proprietor's Association. It is perfectly obvious to us that no reliance can be placed on the statements submitted by them . . . In our view there is not sufficient data available to form a definite opinion one way or the other as to the effect of slate dust on the internal organs, although our members are convinced that it is chiefly responsible for bringing on some forms of Phthisis [Tuberculosis]."*

By the 1950's, however, according to J. G. Isherwood in <u>Slate From Blaenau Ffestiniog</u> (1988), "the quarries had been decimated. . . All were pale shadows of their former selves, their workforces ravaged by the effects of silicosis, a disease ignored for too long by many quarry owners until the evidence was beyond doubt; the most experienced men had received the greatest exposure to the killing dust and so the dearth of experienced miners and rockmen increased."

The struggle between the Welsh quarry workers and the quarry owners was bitter at times. The men who dared to complain about the working conditions could simply be fired from their jobs and possibly evicted from their homes, as quarry owners were often both employers and landlords.

Despite the tough working conditions, a principal complaint among the workers concerned the corruption, bribery, favoritism, and discrimination which seemed to prevail among the quarry officials, who would grant a good rock face (a "bargain") to a worker of similar religious or political bent, while poor rock faces went to others. This forced the men to form unions and to take their complaints to the management by union committee in order to protect the individuals with grievances from losing their jobs.

The first union attempt occurred at Penrhyn Quarry in 1865 where 1800 men joined, and soon lost their employment as a result, until the union was broken. The union organizers were

subsequently fired. This attempt at unionizing caused quarry owner Lord Penrhyn to issue a statement on December 2, 1865, advising the men *"to consider well before listening to agitators,"* and offering *"a word of caution to avoid having anything to do with such a movement as a trade union in future, as on the very first rumor of such a state of feeling, he will immediately close the quarry, and only reopen it and his cottages to those men who declare themselves averse to any such scheme as a trade union."* The closing of the quarry and the men's cottages would subject the men to extreme hardship, especially in the winter months, while Lord Penrhyn could simply wait it out in the Penrhyn Castle, relying on his wealth and servants to continue an opulent lifestyle. A saying eventually arose, attributable to Lord Penrhyn's acquisition of land holdings, *"Steal a sheep, they hang you, steal a mountain, they make you a lord."*

In 1870, over 80 Penrhyn quarrymen, among the best in the quarry in character and skill, were fired without reason, although it was widely believed that the men were fired for political reasons. A letter written by an apparent mediator in 1874, during another strike, perhaps sums up the situation succinctly: *"Over 2,200 quarrymen are out on strike and about 300 old men have been 'turned out' by Lord Penrhyn for whom some provision must be made to keep them from the Union House. Some of these old men have been working at Lord Penrhyn's quarries for 40, 50, and 60 years,*

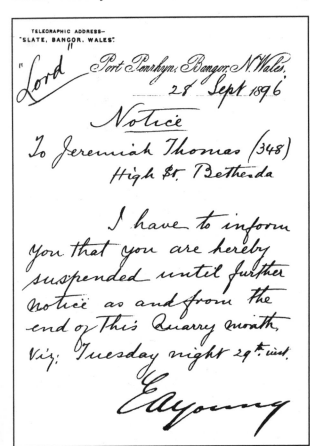

Figure 4.15: An old section of the Penrhyn Quarry as it looks today. The pit has filled in with water, but the terraces cut into the mountainside are clearly visible.

(Photo by the author)

and some of them even more than this; and simply because the bulk of the quarrymen resolved to ask for an increase of wages, these poor innocent old men were being cruelly punished by being thrown upon the Parish for any provision Lord Penrhyn was ready to make for them. This certainly is not charity!"

A more successful, broader union formed in 1874 at the Dinorwic Quarry, attempting to unionize all the slate quarries in North Wales, and was met with the uniform hostility of the quarry owners. The owner of the Dinorwic Quarry attempted to exclude all unionists from the quarry, leading to a five week walk-out until the union was recognized. Another strike occurred at Dinorwic in 1885-6, as a result of complaints by the workers of favoritism. *"We respectfully invite you to examine the grounds for the following complaints; and if you approve of it we shall be ready to appear before you to discuss this, and to enter into particulars if required. We complain that at present, and for some time past, we are not treated as workmen, but that the question of our politics and religion are allowed to affect the question of our work and our wage. In this matter we maintain that clear and offensive favoritism is shewn. . . Let honest work be paid for and not the Political or Religious creed of any man.. . ."* Shortly thereafter, it was agreed that no worker would be discriminated against because of religion or politics, after which Dinorwic experienced a considerable period of peace.

Penrhyn Quarry was not so lucky. Lord Penrhyn refused to recognize the right of union committees to take up grievances on behalf of the workers. A lengthy strike followed in 1896, but work eventually resumed without any grievances being settled. Worker discontent culminated in a three year strike at Penrhyn from 1900-1903, having a disastrous effect on the local town, Bethesda. The military was called in to keep order as thousands of quarry workers and their fam-

Figure 4.16: Meredith Tanyard Cottage in Dollgellau, Wales, an old cottage with a restored roof.

(Photo by the author)

ilies were threatened with starvation. The specter of "Socialist experiments" was raised in the Tory press, depicting the work of the unions to be a creeping menace. Lord Penrhyn adamantly refused to give in to the demand that he recognize union grievance committees, and the strike eventually failed. The union carried on, however, and Lord Penrhyn eventually became *"The most hated man in the world,"* according to a worker at the Welsh Slate Museum, who added, *"We Welsh don't like to talk about him."* And that was the sentiment still in 1996, nearly a century later!

The British government did pass the Quarry Fencing Act in 1887, followed by the Quarries Regulation Act in 1889, and the 1894 Quarries Act, initiating government inspection of quarries and quarry machinery. In addition, the Workman's Compensation Acts of 1880 and 1911 provided greater protection for the workers.

Otherwise, workers turned to their church and/or pub for consolation during hard times. In one small mining village there were more than 29 places of worship within three miles of the center. The chapels were the centers of the community, and were used for important meetings as well as for song and worship. On the other hand, pubs provided some competition as meeting places, and in some towns the pubs matched the places of worship, pub for chapel. By 1880, there were 35 pubs in the tiny town of Bethesda alone (near the Penrhyn quarry).

It's no wonder that the lure of America became so strong to many of the Welsh people. In America they could own their own land and their own home with a garden and livestock, and without the landlords meddling about. Letters of Welsh immigrants spoke of the opportunity they found in America. On January 29, 1869, one wrote, *"I am amazed by the efforts made by the Welsh in Wales to get a farm. When one comes vacant, there are hundreds trying to get it. But here you can be your own master without fear of being turned out and you can do what you like with your own*

Figure 4.17:
The author on a self-guided tour of a hotel in Wales.
(Photo by Jeanine Jenkins)

land." On January 2, 1871, another wrote, *"My old friends in Lleyn [North Wales] can have a small holding for themselves [in America] for the money they pay in rent for one year in Wales. One need not fear any notice to leave from any landlord or steward for voting according to conscience. I never met anyone yet who regretted coming to this country, but only many who were sorry they had not come before."*

And so, the Welsh immigrated to America, discovering slate in York County, Pennsylvania, in the 1700's, and establishing the town of Bangor, which became Bangor West, and is today known as Delta, PA, where a large Welsh church graces the main street. In the Lehigh-Northampton district of Pennsylvania the Welsh discovered slate in the 1800's, and once again a town named Bangor sprung up where it still exists today, just a few miles from another town - East Bangor. Both Bangor and Delta, PA, remain as strong centers of Welsh influence in the US. Another Welsh settlement grew up at Wind Gap, PA, in Northampton County, not far from Bangor.

Vermont fared much the same way. The border between Rutland County, VT, and Washington County, NY, became a mecca in the 1850's for Welsh immigrants drawn to the slate quarries. Many small Welsh settlements sprung up, including Fair Haven, Blissville, Poultney, South Poultney, Pawlet, and West Pawlet, all in Vermont, and Granville and West Granville in New York. One strictly Welsh organization from the region, the Poultney Welsh Male Chorus, was still being organized as late as 1939.

Other Welsh immigrants were drawn to the Virginia slate regions, concentrating in Arvonia, a town named after Caernarvon, Wales. Others were drawn to the Rockmart and Fairmount areas of Georgia. Welsh emigration to America reached its peak by 1900 with nearly 94,000 total immigrants and over 170,000 children of Welsh immigrants present by that time (29,868 total immigrants by 1850; 45,763 by 1860; 74,533 by 1870; 83,302 by 1880; 100,079 by 1890; 93,744 by 1900, 82,488 by 1910; and 67,066 by 1920). The significance of this emigration is drawn into perspective when we realize that the entire population of Caernarvonshire in 1871 was only 106,000, and in Merioneth only 46,000, while the population of Wales in its entirety in 1871 was only 1.2 million. This means that nearly 7% of the entire population of Wales had emigrated to America by 1870, or in today's terms, that would be like 18 million Americans emigrating to (for example) China.

Pennsylvania, Ohio, and New York attracted over one third of the Welsh immigrants and their children, while Illinois, Wisconsin, Iowa, Utah, Kansas, Colorado, Indiana, and California attracted lesser quantities, in that order. Welsh were most numerous in Pennsylvania in the coal mining and steel areas such as Pittsburgh, as well as in the slate areas already mentioned. The census of 1900 shows the Welsh most concentrated in Pennsylvania in the Wilkes Barre-Scranton area, Pittsburgh, and other coal mining areas as well as slate regions.

The Welsh quarry workers who stayed in Wales continued to put in a good day's work, although the quarry managers could never get enough out of them. In 1913, for example, each Welsh mine worker averaged 33 tons of finished slate per year, and each quarry worker averaged 32 tons. Despite the workers' long days, the manager of the Penrhyn Quarry insisted that they only worked seven and a half hours a day *"after deducting the meal hour. And then there is the time when the men are not working, such as blasting time, and if you add to this the loss of time such as holidays, attending funerals, hay harvest, and rough weather, the actual working time becomes small."* Note that he even complained about the men attending funerals! With management like that, it's easy to see why the workers wanted to leave the country.

Figure 4.18: ▼ **A sign that speaks for itself.**
(Photo by author)

▶ **The old post office in Tintagel, Cornwall, England, now a property of the National Trust. Note the "drip-stones" on the chimney bottom, built in to divert water from the joint between the chimney and the roof.**
(Photo by Dave Starkie)

Figure 4.19: ▲ The coastal town of Barmouth, Wales, is a study in slate and stone.
▼ A typical Welsh farmhouse and outbuildings. Note "tingled" repairs on outbuilding - lead strips to hold slipping slates. ▶ New, random-style slates lay propped against a cottage to be re-roofed in Spain.
(Top photo by the author, bottom and opposite by Dave Starkie)

<u>References Chapter Four - Wales</u>

•Brigden, John, (date unknown), *"Turning Stone into Bread,"* (publisher unknown). This article was on file at the Buckingham-Virginia Slate Quarry in Arvonia, Virginia.

•Carrington, Douglas C., and Rushworth, T. F., (date unknown), <u>"Slates to Velinheli - The Railways and Tramways of Dinorwic Slate Quarries Llanberis,"</u> found in Green Mountain College, Poultney, Vermont.

• *"Chwareli a Chwarelwyr,"* (Quarries and Quarrymen), 1974, Gwynedd Archives Series, (A booklet to accompany an exhibition prepared to celebrate the centenary of the founding of the North Wales Quarrymen's Union), pp. 47-51.

•Hartmann, George E., (1967), <u>Americans from Wales,</u> The Christopher Publishing House, Boston, MA, pp. 86-91.

•Holmes, Alan (1986), <u>Slates from Abergynolwyn,</u> (The Story of Bryneglwys Slate Quarry), Gwynedd Archive Services, Caernarfon, Gwynedd, Wales, pp 113-114.

•Isherwood, Graham (1982); Cwmorthin Slate Quarry; Revised edition published by Adit Publications, Towerside, Pant-y-Buarth, Gwernaffield, Mold, Clwyd, CH7 5ER, UK.

•Isherwood, J. G., (1988); <u>Slate From Blaenau Ffestiniog;</u> AB Publishing, 33 Cannock St., Leicester LE4 7HR England.

•Lindsay, Jean, 1974, <u>A History of the North Wales Slate Industry,</u> David and Charles, Newton Abbot, London, pp.11-27, 246-61, 286-295, 324-325.

•Encyclopedia Britannica, (1965), Volume 23, p. 296 (population of Caernarvonshire and Merioneth in 1871).

•Lewis, M. J. T., and Denton, J. H., (1874), <u>Rhosydd Slate Quarry,</u> The Cottage Press, Shrewsbury, England, P. 98.

•McKinney, Margot, 1976, *"The Welsh Heritage of the Slatebelt,"* produced at Green Mountain College under a grant from HEW, Poultney, Vermont, Journal Press, Inc.

•Richards, Alan John (1994), <u>Slate Quarrying at Corris;</u> Gwasg Carreg Gwalch, Llanrwst, Gwynedd, Wales.

•Richards, Alan John (1995); <u>Slate Quarrying in Wales;</u> Gwasg Carreg Gwalch, Iard yr Orsaf, Llanrwst, Gwynedd, Wales.

• *The Slate Industry of North Wales, Unit 4;* A collection of historical documents available at the Llechwedd Slate Quarry historical site at Blaenau Ffestiniog, North Wales.

•Williams, Merfyn (1991); <u>The Slate Industry,</u> C. J. Thomas and Sons, (Haverfordwest) Ltd., Press Buildings, Merlins Bridge, Haverfordwest, Dyfed SA61 1XF, UK.

•Williams, M. C., and Lewis, M. J. T., 1989; <u>Gwydir Slate Quarries,</u> Snowdonia National Park Centre, Plas Tan y Bwlch, Maentwrog, Blaenau Ffestiniog, Gwynedd LL41 3YU, Wales.

·*·EUREKA·*·
SLATE COMPANY,

MANUFACTURERS OF ALL COLORS OF SLATE,

THE ONLY

EUREKA!
UNFADING

HUGH G. HUGHES,   R. WYNNE ROBERTS,

POULTNEY,   18 Little Tower,

VERMONT.   LONDON, En

ESTABLISHED 1852.

EUREKA SLATE QUARRIES

Are now the Oldest existing in the State.   These Quarries are now produc
the best, and, in fact, the only unfading Green Slate in the world,
and have gained this reputation throughout the States
and Foreign Countries as well.

POULTNEY, - - VERMONT.

Chapter Five
Vermont - New York

The "Slate Valley" of Vermont lies mainly in Rutland County, Vermont, but also straddles the state border to include some of Washington County, New York. The 24 mile long, six mile wide valley extends from Granville, NY, and West Pawlet, VT, north to Fair Haven, VT. The roof slates from this area are well known both for their durability and for their variety of colors, ranging from deep red to solid purple, purple with green flecks or streaks, gray-green, solid green, "sea green," gray with black streaks, gray-black, and black.

This differs greatly from the slates of Monson, Maine, eastern Pennsylvania, Peach Bottom, PA, or Buckingham, VA, which are all black. Although some green, purple and blue slate may be found in eastern Canada, at Burgoyne's Cove in Newfoundland, as well as in small deposits in Arkansas, Tennessee, Georgia and Utah, the only commercial source of colored slate in the United States, of any significance, lies in this area straddling the border between New York and Vermont.

The bulk of the green and the purple slate is found on the Vermont side of the valley, and all of the red slate is found on the New York side. It can be fairly said that all of the true red slate in America comes from Washington County, NY. First discovered in the 1850's near Green Pond in Hebron, in an area now known as Slateville, red slate is now one of the most durable (and most expensive) slates available anywhere.

The slate quarried from the Vermont side of the valley is primarily green, "sea green" (green that develops a mottled reddish cast over the years), and purple, but also includes some Vermont black and gray.

The first quarrying in the Slate Valley took place in 1839 near Fair Haven, Vermont. A fellow by the name of Colonel Alonson Allen started the quarry at a place called Scotch Hill, just north of Fair Haven. Allen was born in Bristol, Vermont, on August 22, 1800, and he settled in Fair Haven at the age of 36 as a proprietor of a small general store. He started his slate quarry with Caleb B. Ranney.

For a few years, Colonel Allen dabbled experimentally with the slate, then in 1845 he went into the business of making school slates. Hand-held pieces of slate were used by school children instead of paper writing tablets in the mid-1800's, and much of the slate quarried at that time was used to produce these school slates. In fact, slate was also used to make slate *pencils*, which were made entirely of solid slate (no wood, no graphite). A slate pencil was about three or four inches long and a quarter inch in diameter, pointed at one end, and used for marking school slates, much the same as a pencil is used to mark a piece of paper today. When a sharp piece of slate is dragged across a flat piece of slate, a white mark remains on the flat piece

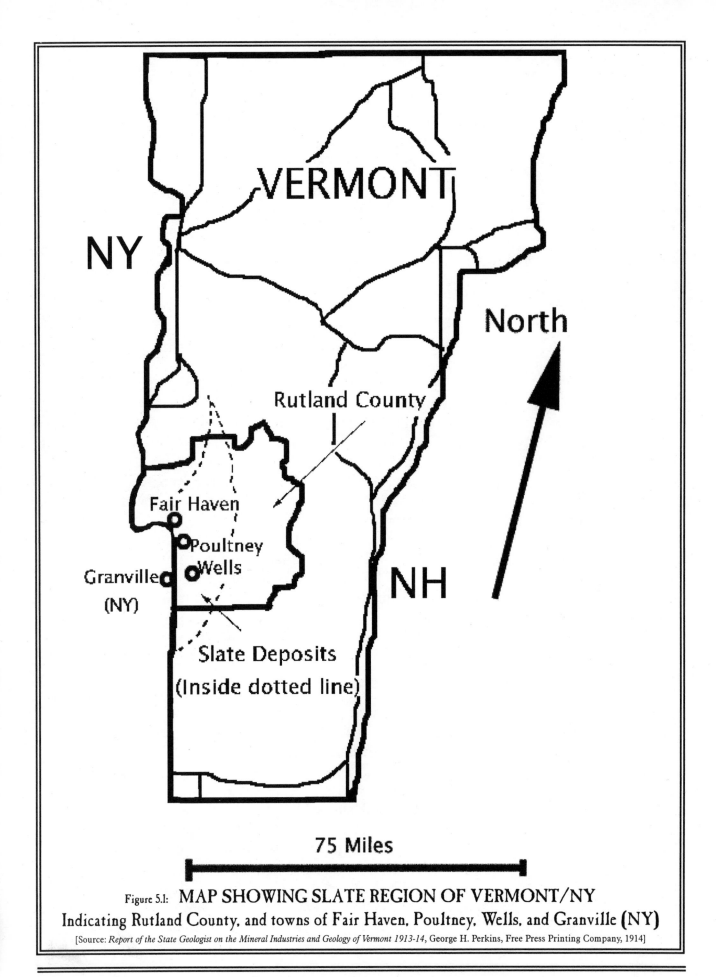

75 Miles

Figure 5.1: MAP SHOWING SLATE REGION OF VERMONT/NY
Indicating Rutland County, and towns of Fair Haven, Poultney, Wells, and Granville (NY)
[Source: *Report of the State Geologist on the Mineral Industries and Geology of Vermont 1913-14*, George H. Perkins, Free Press Printing Company, 1914]

Figure 5.2: This Vermont "sea green" slate roof near Fair Haven, Vermont on Scotch Hill Road, near the first slate quarry in Vermont, has withstood the test of time and is still in good condition after 145 years. If cared for, the roof still has many decades of life left in it. [Photo by the author]

which is easily rubbed off by a piece of cloth or even a child's hand. Chalk is not needed when one has a slate pencil. And so, thanks to the quarrying of slate, many schools were provided with both "tablets" and writing utensils for the children. In addition, slate "blackboards," hung on walls, were used by teachers and students. Today we have sayings such as "wipe your slate clean," or have a "clean slate," or events are "slated," or a "slate" of candidates. These are all carry-overs from the days when slates were used as writing tablets.

Colonel Allen didn't stay in the school slate business for long, however, because he found he could make more money making roofing slates, which were about the same size and thickness as school slates. He began roofing slate production in 1847, just two years after starting the manufacture of school slate, then he abandoned his school slate production a year later, in 1848.

The first slate roof covered by Vermont slate was done by Colonel Allen in 1848, but the farmer who owned the roof was a skeptic. He insisted that Colonel Allen wait a full year for payment in order to determine if the roof would collapse under the weight of the slate. If the roof did collapse, Colonel Allen was to receive no pay, and he was instead to pay for all damages. It turns out that the roof passed the test and lasted quite a bit longer than the skeptic's one-year trial period. In fact, it still exists in good condition today, one hundred and fifty years later. The photo on this page (Fig. 5.2) shows a house on Scotch Hill Road with a sea green roof installed in

1851, a scant 3 years after Colonel Allen installed the first slate roof in Vermont. The photo was taken in 1993 and clearly shows that a sea green roof will last for centuries if cared for.

Allen also served as state Senator in 1842-3, and as an assistant judge of the county court in the early 1860's. In 1886, eight years after Allen's death, one Rutland County historical account stated, *"but for his boldness and courage, to this hour not one slate would have been shaped from Fair Haven to Salem."*

Colonel Allen opened a floodgate in 1848, after which slate quarries began popping up everywhere, following his example. The "Eagle Quarry" was opened just outside of Poultney, Vermont, in 1848; then in 1851 another quarry was opened on a farm owned by Daniel Hooker three miles north of Poultney village. Hooker and Son quarried a few slate in 1851, then a few more in 1852 as they were able to sell them to their neighbors. By 1854 they had a full fledged operation going, manufacturing mainly roofing slate and eventually employing as many as 60 men at one time. Some of their slates were of such high quality they were even exported to England. Then Welshmen began opening quarries right and left. By 1885, forty-six years after Allen started the first quarry in Vermont, 67 slate quarries were operating in Rutland County.

By 1871 slates were also being manufactured for mantels, billiard tables, hearths, table tops, blackboards, floor tiles, door steps and other articles. Some slate was "marbleized," which involved a process of painting the slate black, heating it to 175°F, then dipping it in water that had paint floating on the surface. The slate was baked again, varnished, baked again, polished, and baked a fourth time until the final product looked incredibly similar to marble. An 1875 description of the process states, *"The ingenious process of marbleizing is one of the recent inventions. By a certain chemical process the surface of the slate, after it receives a polish, is converted into the exact semblance of the most beautiful of the foreign and domestic marbles, and also made to imitate rosewood, mahogany and ash, and so exactly that the most experienced would be puzzled to detect the difference from sight."*

One typical quarry by this time was being worked to a depth of 150 feet from the surface, all dug without today's modern machinery. From this pit opening six tunnels were dug into the

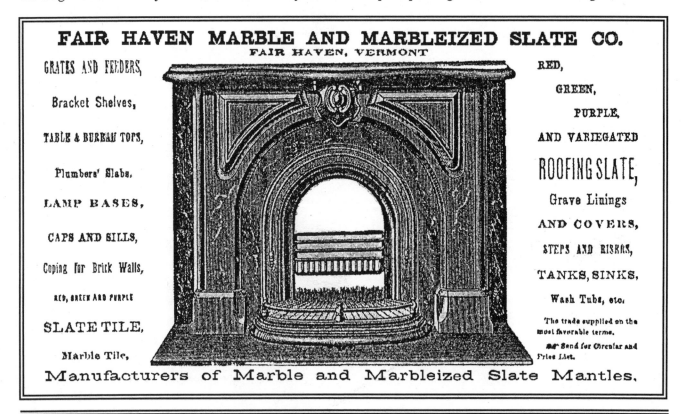

hillside to a distance of 600 feet, the slate being removed on railways operated by steam. The slate deposits in this area were considered *"inexhaustible,"* and still exist in huge quantities today.

By 1872, over 11,000 tons of roofing slate were being shipped annually from Poultney, Vermont, amounting to about 35,000 squares of roofing (enough to cover about 3,000 average houses). The cost per square was about $4.50. The cost today (1997) is about $350.00.

Visitors to the Slate Valley today can camp at Vermont's beautiful Half Moon Pond State Park, and see the old Scotch Hill quarries along Scotch Hill Road, which travels between the park and the lovely town of Fair Haven. While in the area be sure to visit the Slate Valley Museum, 17 Water Street, Granville, New York 12382 (Ph: 518-642-1417); or visit the web site at http://www.slatevalleymuseum.org.

Eleanor Evans McMorrow was born in Wales near Caernarvon, and came to Vermont with her parents in 1924, at the age of seven. She tells of Caernarvon Street in Fair Haven, an example of the Welsh influence in the development of this area of the United States. Almost all of the Welsh in the Vermont/New York slate region were from North Wales where the Welsh slate quarries are located, she says, while those from South Wales, who mostly worked coal, immigrated to Pennsylvania to work the coal mines.

Eleanor's Welsh maternal grandparents had lived in Vermont in the mid-1800's but they eventually returned to Wales. Her maternal uncles had also moved from Wales to Poultney, Vermont, and in doing so influenced Eleanor's parents to make the great Atlantic journey themselves. Her father started working in slate in Wales when he was eleven, not an uncommon age for Welsh lads to start in the "pit," as they called it. He became quite an expert at slate, and helped many a person who wanted to start a slate quarry. *"He could look at the rock on the top and tell them how to start whatever they did to the slate. They have to get it in a certain vein or something*

MORE BORING STUFF

Figure 5.3: ANALYSIS OF VERMONT/NY SLATE (%)

	Sea Green	Unfading Green	Purple	Red
Silica	65.02	64.71	62.37	73.93
Protoxide of iron	5.44	5.44	4.21	1.74
Peroxide of iron	2.99	7.23	7.66	10.17
Alumina	16.02	7.84	13.40	5.16
Manganese Oxide	0.31	0.30	0.20	0.10
CA Carbonate	1.38	3.00	2.50	1.25
CA Sulphate	1.31	1.55	0.16	1.06
Phosphoric acid	trace	trace	trace	trace
Alkalies (NA)	4.16	6.92	7.20	3.92
Water	1.37	1.38	1.50	1.24
Magnesia	2.00	1.63	0.90	1.43

[From <u>History of Rutland County, Vermont</u>, edited by H. P. Smith and W. S. Rann, D. Mason and Co., Syracuse, NY, 1886, p. 198. Data attributed to Professor J. Francis Williams, Rensselaer Polytechnic Institute, Troy, NY.]

Figure 5.4:

▲ James Kelly of Wells, Vermont, is one of the few remaining blacksmiths still making tools for the slate industry. ▼ He starts with the iron bar stock at left, rough forges it into a blank for a splitting chisel (center), then hammers and forges it into the finished product (right).

(Photos by the author)

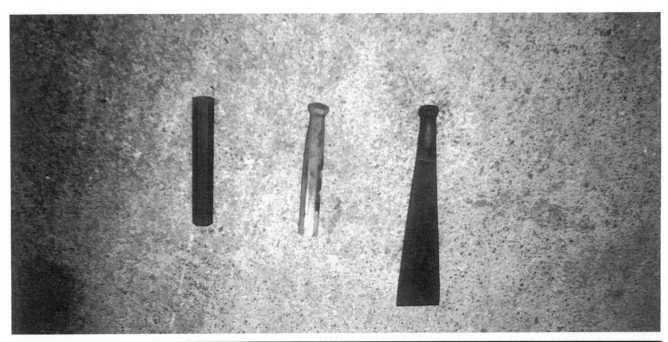

and he could do that," she said. He worked in the pit in Vermont until about a year before he died.

Once in Vermont, Eleanor found that her church was Welsh and all the services were in Welsh. She went to church seven days a week, as it was the center of social life. *"The Welsh sang all the time,"* according to Eleanor, whose dad, uncle and brother sang in choirs and choruses.

The slate they quarried was roofing slate. *"This whole area is slate. There are many, many, many colors, beautiful colors. Middle Granville [NY] has a beautiful red. And then there's purple and black, gray. Then there's one that's sort of mottled green and purple. It's awfully pretty."*

Life in the quarries could be very hard, according to Eleanor. *"I'm very happy I don't have anyone in the slate quarries anymore. We were brought up to get your bed made, your dishes done and your floor swept, in case somebody gets hurt and they bring them home from the quarry. That was the way you were brought up. My father was hurt badly. He was unconscious for 21 days. They operated on his brain. He was home for nine years after it. He had to learn to write and everything over again, learn to talk."*

"They have these big blocks they send up. It was a bad day, it was sleeting. A block came out of the chain and hit him in the head. My father got $9.00 a week in compensation. I don't know how they made ends meet, if they did."

There were always injuries, says Eleanor, who tells of broken arms and broken legs. *"Somebody was always getting killed, it was terrible. More so here than in Wales, I think, from what I've heard them tell. Over there the quarries were mined in shelves. And here, right down."*

The huge dumps of waste slate around the quarries [Fig. 5.21] were also hazardous, according to Eleanor. *"The man that was boarding with us, ten days after my father was hurt, this guy made him go and work in the same spot my father was in. These dumps get icy, they slide. And it slid and killed him. He was knocked right off the ledge and into the quarry. He was killed ten days after my father was hurt. Today they wouldn't be able to allow that because of your safety codes. But they didn't have safety codes back then, in 1938."*

SEA GREEN SLATE

Vermont is a beautiful state, and it has produced perhaps more hard slate than any other. "Sea green" (now known as semi-weathering gray or green) slate roofs can be found at almost any location in New England served by railroads at the time the slate was quarried, although many areas not served by railroads at that time do not have slate roofs today, and never did. For example, just a few miles east of the slate quarrying region of Vermont, over the Green Mountains, slate roofs are hard to find, as there was no railroad access to that area during the heyday of slate quarrying. On the other hand, Vermont's sea green slate can be found hundreds of miles away to the south and west where the rail lines routinely hauled the slate to Pittsburgh, and farther west to Ohio and beyond.

Sea green slate is easily identifiable because the roof will appear light gray with reddish slates scattered throughout (it only appears "sea green" when freshly quarried). The reddish color is also called buff, tan, pink, orange, rust, or brown, but since it is caused by the oxidation of iron layered into the slate itself, we may as well call it reddish. People in the slate industry do not like to use the term "red" when describing the mottled color of sea green slate because true "red" slate, quarried in New York near Granville, is a deep, solid (not mottled) red similar in color to red brick, and does not resemble sea green slate at all.

Thanks to Margot McKinney, a reference librarian at the Green Mountain College in Vermont in 1976, a direct history of the sea green quarries has been put to record. We are even more grateful to Owen L. Williams, the author of the history, who, at the age of 89, in 1969, penned his fifth revision of "The History of the Sea Green Quarries," on a six inch by nine inch dime store tablet. Margot McKinney found and typed the contents of the tablet after Mr. Williams death in 1975 in Wells, Vermont, and recorded it in the school records of 1976, where it found its final resting place in a manila envelope in a drawer in an upstairs room in the school library. Here it is now excerpted and paraphrased to provide a first-hand account of the sea green quarries *"by an old slate maker of over sixty years experience. . ."*

Owen Williams was born in 1880 in Caernarvonshire, Wales, and emigrated to the United States in 1901. Twenty-six years later he sat in the Vermont Legislature, despite the fact that he never had any formal education beyond the seventh grade.

He wrote the *History of the Sea Green Quarries* five times, each time a revision of the previous version, all handwritten on 6 x 9 inch tablets. The final version, written in 1969, bore the words, *"This is the last I wrote."*

The sea green quarries were located in the southwest part of Rutland County, Vermont, specifically near Fair Haven, Pawlet, Poultney and Wells. The slate deposit covered about 10 miles or more, and varied in width from a few hundred feet to half a mile or so.

According to Mr. Williams, various records show the opening of the first sea green quarry to have taken place in the late 1860's or early 1870's, although other records apparently not available to Mr. Williams indicate that slate was being quarried several years earlier than this. This is best evidenced by Figure 5.2, showing a sea green roof dated 1851. Perhaps Mr. Williams' reminiscences did not include the Fair Haven area of Vermont (not far from Poultney) and Scotch Hill, where the first Vermont slate quarry was opened.

The sea-green slate deposits were, of course, below the surface of the earth, and an excavation had to be sunk into the ground in order to access the slate and begin to open a quarry. So it was necessary for the quarryman to put one or several holes down and blast out a section of overlaying rock in order to reach the slate and determine whether the slate was of adequate quality to continue. If the slate looked promising, then more holes were drilled at an angle corresponding to the layers of the slate, and blasting was continued until the quarryman has cut a *"butt and a fireside."* *"These terms,"* states Mr. Williams, *"are used by quarrymen. Please bear in mind that these terms that I use in this history prevailed in the early days with nothing but hand labor."*

When the overlaying rock was covered by soil, sand or clay, a team of horses with a scoop shovel would drag a trench across the vein to expose the rock. Then, using the same technique, the trench was widened until the rock could be reached and blasted. The overlying rocky slate

---

Figure 5.5:   **SEA GREEN SLATE**

Since there are so many century-old sea green slate roofs in the United States, it's important to know that 100 year old sea green slates are usually still very durable, and may yet last another 100 years. Although most sea green slate is classified as hard, some sea green quarries produced a softer slate which may appear brownish and flaking with age. These may be relatively soft and may only last 150 years.

Strength:	11.040 lbs/square inch (modulus of rupture)
Porosity:	.220% water absorbed in 24 hours
Corrodibility:	.722% of weight lost in acid solution for 63 hours
Silica:	67.76%
Carbon:	None
Alumina:	14.12%
Ferrous Oxide:	4.71%
Plus many other misc. minerals	(Source: U. S. Geological Survey, date unknown)

was often not solid enough *"as quarrymen say to make roofing slate of good quality and durability"* to a depth of ten to fifteen feet, although in some years the top rock was in demand to make antique-looking ("freak") slate.

In those early days of slate quarrying, it was all hand labor. Hand shovels were used to take care of the rubbish, and elm or oak plank boxes, braced with strap iron and measuring four feet wide and five feet long, were used for moving slate, and for dumping scrap (Figures 5.12, 5.13, 5.14). At the turn of the century (1900) self-dumping boxes came into use, made of iron. Other tools included both double and single-hand hammers (requiring one or two hands, of course), crow bars, drills, wedges, chisels, gouges, rope, black powder and fuse. Dynamite was used on poor rock, but never on good slate rock *"on account of the shattering effect."*

Drilling was all done by hand. One man held a metal rod fitted with a drill bit on its end, while another man (or two) struck the rod with a double handed hammer. The man holding the drill turned it a little after each strike, boring a hole into the rock suitable for holding explosive black powder. This drilling process was very dangerous and required the strict attention of the three men; otherwise, the man holding the drill would be seriously, even permanently, injured by a blow from one of the heavy hammers.

The fuse was then placed into the hole, followed by the black powder. After the blast, the slate had to be *"cleaved"* (split apart like a deck of cards). The cleaves were made ten or twelve inches in thickness, and if a piece of stone was too large to handle, it was *plugged* (split on a plane perpendicular to the layers, or broken in two). Plugging again involved drilling a hole by hand, although in this case only one man drilled the hole using a short drill (about two feet long) and a single-hand hammer. This hole was not to be used for explosive powder, however, but instead a *plug and feathers* was inserted into the hole to force the block of slate to split in two.

A "plug and feathers" is a set of tools consisting of two thin steel sleeves (the feathers) that slide down either side of the hole, after which a tapered wedge (the plug) is forced between the feathers with a hammer. The tapered wedge caused the block of slate to split in two, while the feathers provide a smooth surface allowing the wedge to be driven into the hole. This way, slate can be split along a plane perpendicular to the layering of the stone. Figure 5.6 below shows

Figure 5.6
SLATE MAKER'S TOOLS

1) <u>Chisel</u> - For splitting large blocks down to workable size.
2) <u>Mallets</u> - For pounding chisels.
3) <u>Splitting chisels</u> - For splitting out individual roof slate.
4) <u>Gouge</u> - For making groove in slate block.
5) <u>Plug and feathers</u> - For inserting into hole to split block of slate.
6) <u>Hand hammer</u> - For tapping plug and feathers into place.

a plug and feathers, some chisels, both one and two hand hammers, and other tools used by slate makers.

Instead of a drill, a *"jumper"* could be used to plug a piece of slate. A jumper is a steel rod about six feet long with a small, but heavy ball of iron welded into it about fifteen inches from the end to provide additional weight in the drilling process. This tool required a good deal of experience in order to drill a hole suitable for a plug.

Another tool frequently used by quarrymen was the *gouge*. According to Mr. Williams, the gouge was made of 5/8 inch steel, about 10 inches long (when new) with a point suitable to cut across the endgrain of a stone, in order to *sculp* it - make a cut the length of the stone *perpendicular* to the obvious splitting (cleavage) plane (see Figure 2.4, page 23). If the resulting *ditch* on the end of the stone was cut smooth and straight, the chiseled split would be very good and made in the direction desired, but if the ditch was not made right and straight, the result would be very unsatisfactory, and if so would require more labor and a waste of valuable stock.

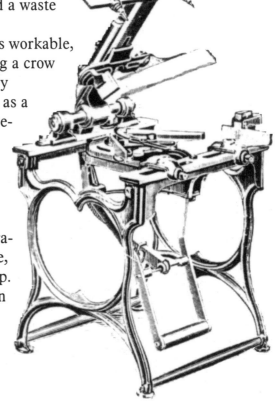

Figure 5.7
Ruggles Foot Powered Drill and Slate Punch

Once the stone had been split to a size that was workable, it had to be hoisted out of the pit. This was done using a crow bar to manipulate the block while a chain was securely wrapped around it. Here was another hazardous task, as a stone could slip from the chain and injure or kill someone as it was being hoisted out of the quarry. The chain was about 12 feet long and made of half inch wire, with links two and a half inches across. *"Sometimes that rockman had to use a rope to hang on to in order to put the chain around the stone secure."*

Slate making itself was divided into three operations: block cutting, splitting, and trimming. As a rule, says Mr. Williams, there were three men in each group. The function of the block cutter was very important in the production of roofing slate, requiring a great deal of experience. It was a job that could wear a man down very quickly unless he knew what he was doing. Blocks came out of the pit in irregular shapes and the block cutter had to make the blocks suitable for splitters to work. If the block needed to be split using a plug and feathers, the block cutter needed

**Figure 5.8 The "Lightning" Slate Dresser --
punches nail holes in slate.**

to measure for the proper place to put the hole.

If the stone needed to be sculpted, then the block cutter had to use his gouge to cut a groove or ditch in the end or side of the stone. From this groove, the block cutter cut into the block with a chisel in the direction he wanted. Then he had to cleave the block (split it along the obvious splitting plane) to make the pieces light enough to carry into the splitting shanty. This was all done outside.

A person referred to as the *"splitter"* sat in the splitting shanty. His tools consisted of a low stool to sit on, a wooden mallet with iron rings at both ends, and two or three thin, flat iron chisels about three inches wide at the wide end, which were used to split out the individual pieces of roofing slate. The splitter (or "slitter" in Wales), sat on the stool and leaned the slate block against his leg with the smoothest edge upward. The block had now been split to a thickness of about one and a half inches by the block cutter. The splitter then split the block along the cleavage plane into two equal halves with the hammer and chisel. A good piece of slate will split right into two halves with a good tap or two of the hammer. Each of the halves was again split in half, and so on, until the desired thickness of about 3/16 inch was reached. Naturally, some of the finished slate would be thicker or thinner, but for the most part, the finished slates were amazingly uniform in thickness. Again, this craft required a lot of experience; some consider it an art. The splitter kept the unsplit blocks piled on his left side, and the *"chips"* (untrimmed slates after splitting) piled on his right on a small bench within easy reach of the trimmer.

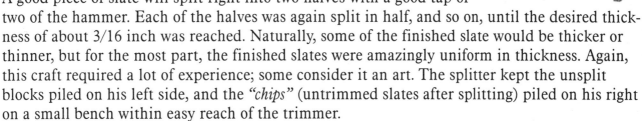

A person known as a *trimmer* then took the chips one at a time and trimmed them to make the largest possible size roof slate. The trimming was done by a trimming machine run by a foot treadle, so that the trimmer had to hold and position the slate with his hands while he operated the foot treadle at the same time (Figure 5.10, at right). The maximum *standard* size of a roof slate in the USA is 14" wide and 24" long, and the sizes range down to about 6" wide by 10" long. In between these sizes are 12", 11", 10", 9", and 8" wide slates of different lengths. Although it was a very arduous job, some men could trim 10 to 15 squares of roofing slate in a ten hour day, amounting to about four to five tons of rock. The trimmers also had to shovel their rubbish as well

Figure 5.10
Ruggles Original Slate Trimmer

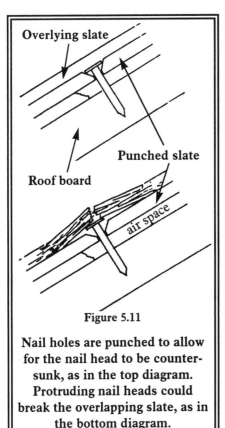

Figure 5.11

Nail holes are punched to allow for the nail head to be counter-sunk, as in the top diagram. Protruding nail heads could break the overlapping slate, as in the bottom diagram.

as the splitter's, and sometimes help the block cutter with his work.

After trimming, nail holes were punched in the slate. This was done by placing each slate, one at a time, on a slate punch. A foot pedal operated the machine which caused two sharp metal punches to poke nail holes in the slate about one fourth to one third of the way down from the top and about an inch and a quarter to two inches from the sides (Figures 5.7, 5.8, 5.9). The punch breaks out the back of the slate when it pokes the holes through, and this breakage was advantageous as it allowed for the head of the nail to be countersunk into the slate (Figure 5.11, at left). This kept the nailhead from rubbing on the slate overlapping above (after being nailed to a roof) and prolonged the life of the roof.

A foreman kept a record of the production of each slate shanty, resulting inevitably in an atmosphere of competition and rivalry between the shanty crews when their individual rates of production were compared. The bosses encouraged the rivalries in order to gain a greater level of production. In some companies, a group of about six men known as a *slate gang* had the job of counting and piling the slate and loading up the teams of horses.

The mornings began with the whole gang in the quarry carrying the previous day's production out into the slate yard, before the pitman went into the pit. Because of the great weight of slate, this was heavy work, and the custom prevailed that there was a ten minute break after the slate was carried out. The slate was then loaded onto wagons drawn by teams of horses, and taken from the slate yard to railroad cars. Here the slate had to be packed into the cars properly - they were stacked three high on the long edge with wooden lath placed between the rows. An article in the Granville Sentinel dated March 1890 stated, *"Charles Peck drew 24 and a half squares of 18 by 9 slate on one load from a Rising and Nelson quarry to the station."* This load, weighing nine tons, was probably drawn on a sled, and certainly pulled by horses. During the first quarter of the 20th century, so much slate was being shipped out of Vermont's slate valley that the Delaware and Hudson Railroad ran a freight train called *"The Slate Picker"* primarily to pick up loads of slate.

**Figure 5.12
Ruggles Slate Car**

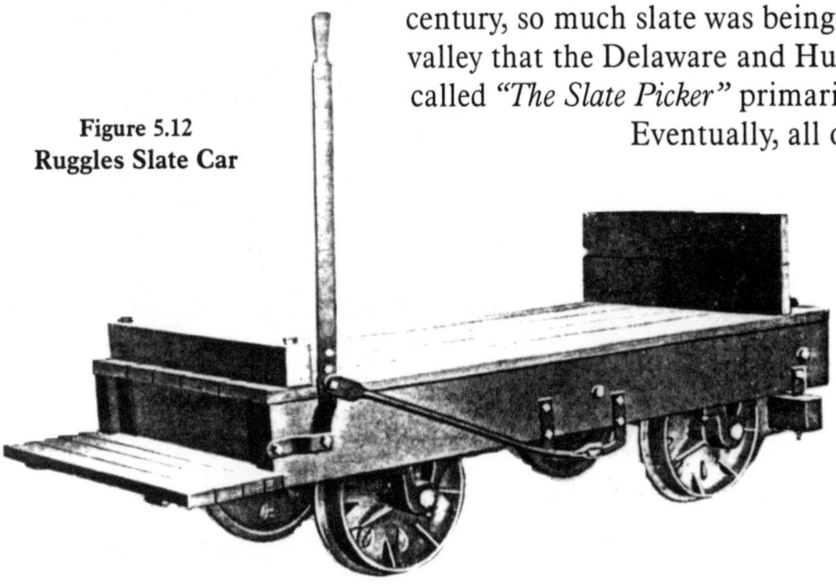

Eventually, all of these slates ended up on someone's home, barn, garage, church, grange hall, carriage house, or other building.

In 1889, the going price for a square of slate was around $3, while the average pay of a quarry worker was about $25 per month. In 1890, one slate company sold 170,000 squares of slate, enough to cover 14,000 roofs, weighing more than 10 million pounds and requiring 3,400 freight cars to haul to market. The peak of slate production in the slate valley occurred in 1927-28 when a square of roof slate went for around $15, compared to about $350 today.

Often a horse provided the power to hoist stone out of the quarry, but as the pit got deeper a steam boiler and engine were used. Eventually, an A-frame supported a cable that extended across the pit, and a *carriage* was hung on the cable (Figure 5.15). The carriage was attached to the engine with a wire rope, enabling the power of the engine to pull the carriage along the cable. The carriage could be brought to a stop by bolting a wooden block on the cable, and this stop was located near where the men needed to unload stone or rubbish boxes. The slate blocks and rubbish boxes were put on a small railcar (Figure 5.12, 5.13, 5.14) - the blocks were rolled to a shanty while the rubbish was rolled to the dump. Eventually, a self-dumping carriage was invented by a blacksmith in Poultney, Vermont, and it came into universal use in slate quarries in a very short time, bringing with it a brand new job (Figures 5.16, 5.17). The self-dumper saved a great deal of human labor as the engineer could stop the carriage anywhere along the cable and set the block down in a convenient location for it to be worked on.

As the pits got deeper they began to fill with water which had to be pumped out. The water was pumped to the boiler house to be used to make steam to power the steam engine.

A very high *mast*, or *stick*, was needed in order to use these carriages (Figure 5.16). The masts varied in length from less than a hundred feet to over two hundred feet tall, depending on the topography of the land. They had to be located four or five hundred feet from the edge of the pit to allow for dumping of the rubbish stone. There was a

Figure 5.13 Wooden Dump Car

Figure 5.14 Wooden Dump Car With Skip

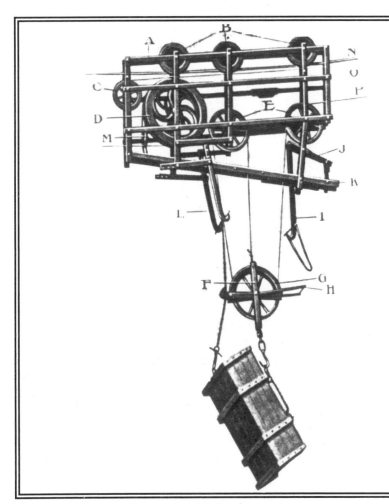

Figure 5.15
AERIAL CARRIER
(Rides on cable between masts)

A. Frame
B. Cable Pulleys
   (Sheaves)
C. Detaining Sheave
D. Brake Wheel
E. Draw Rope Sheave
F. Billy Wheel
G. Billy Wheel Sheave
H. Shifting Bar
I. Shifting Bar Hook
J. Shifting Bar Lever
K. Brake Lever
L. Dumping Hooks
M. Turnbuckles
N. Cable
O. Detaining Rope
P. Draw Rope

Figure 5.16
Masts and Aerial Carriers with Scrap Piles Below

Figure 5.17: Aerial Carrier in Use

demand for qualified men who could raise the mast, which was quite a feat of engineering in those days, considering that the stick may have been 200 feet long. The *"tallest quarry pole in the world"* was 235 feet high erected at Rising and Nelson's Quarry #4 in 1908. A fellow by the name of Henry Vogel of Truthville was a pioneer in this business, and he was so highly regarded that his services were in demand (and he accomplished them even after he became handicapped in one leg).

The sticks as a rule were native green pine. They weren't made of a single solid piece of wood, but were made of two or three poles spliced together to form one piece. The splices were made of eight inch square oak timbers about sixteen or twenty feet long, bolted to the stick. A cable, two guy wires, *"saddles,"* and *"sheaves"* (pulleys) had to be attached to the top of the mast before it went up. All of this together made for quite a heavy mast, and raising it was extremely difficult. One mistake and it could be down in pieces, and the work of a team of men who had toiled for days would be lost. One can still see the old masts standing tall among the scrap slate piles in Vermont today.

The cables were strung high in the air from mast to mast, and the aerial carriers rode on the cables. Despite their height, these carriers could reach into the deepest pit by dropping a box attached to another cable. The slate block or rubbish could then be loaded onto this box and raised up out of the pit, moved along the cable, then let down near the splitting shantys (if a slate block), or dropped into the dump (if rubbish). All this was done by the engineer, who operated the aerial carriers, perhaps three at one time, powered by a steam engine.

Electric power came into the slate valley in 1913. Eventually, air-powered jack-hammers came into use for cutting slate, and could drill a hole in a slate three times as fast as a man could by hand. Around 1919, trucks began to be used to haul slate to the railroads; the first ones had

Figure 5.18: Modern block cutter at U.S. Quarried Slate Co., near Fair Haven, VT, splitting an incoming block using a hammer and a chisel. The block is then taken into the shop and cut roughly to size with a diamond saw, then split into shingles. (Photo by the author.)

hard, solid tires.

The early work forces involved in the sea green quarries were of Irish and Welsh nationalities. These people learned the trade of slate quarrying in Wales and emigrated to the United States, both as skilled workers and common laborers. Toward the end of the nineteenth century, Eastern Europeans began to emigrate to Vermont to work the quarries, and by the turn of the century, the Eastern Europeans came in large numbers. Italians soon joined them in the quarries, and also became very good at rock work. There was only a sprinkling of other nationalities working slate in Vermont.

These four ethnic groups did not mix socially, except perhaps to share a drink at a local bar. Each had different religions, which created the main social barrier as each group's church provided their social center. The Welsh were Protestants and the Irish and Italians were Catholic. The Eastern Europeans had their own church and fraternity.

Throughout the history of the sea green quarries and up until the 1930's there were no health or safety regulations. There were also no public conveniences available to the workers. The men had to provide their own drinking water *any way they could get it, and sometimes under very unsanitary conditions, in some cases melted ice water from some hole under an old dump which would dry up when the hot weather came.*

Slate quarrying was a hazardous occupation due to the cleavage of the stone and the different natural cuts and joints. Undermining was a common cause of accidents, as was carelessness and faulty equipment. The use of black powder and dynamite added to the perilous nature of this trade. There was no workman's compensation, nor were there safety inspections in the early years. The responsibility for safety rested upon the shoulders of the operators and supervisors.

As a result, accidents were frequent, and it was not uncommon for a man to be killed or maimed for life. In 1903, eleven men were killed in a cave-in at the Vermont Slate Co. Quarry located near the Warren Switch section of the sea green quarries. A section of ground where slate was stock-piled and where the part-owner of the quarry, Mr. Williams, happened to be standing, was undermined and collapsed, taking Mr. Williams down and the slate on top of him. In another accident in 1910, five men were killed in a cave-in in the Owens Quarry on Briar Hill, again the result of undermining. One man was found alive the next day after being buried for 24 hours, but died the following night from exposure.

Another incident occurred in the spring of 1916. The quarry was situated such that the engineer who operated the steam engine and dumped the rubbish from the aerial carriers onto the dump pile, could not see the top of the dump. The quarry boss saw a slab of slate on the dump which he thought should have been worked into slate shingles, but the block cutter disagreed. The two of them proceeded to argue over the block while standing in the dump, not hearing the familiar sound of the rubbish box approaching overhead (Figure 5.16 illustrates the probable scene). The engineer dumped the rubbish on them, not aware that the men were standing on the rubbish pile; both men were killed. From then on, the engineer gave two blasts of the quarry whistle as a warning when the rubbish was about to be dumped.

The lack of worker's compensation during those days made for some tough times for families that lost a family head to death or injury. A custom prevailed in which men could pledge money to be taken directly from their pay to benefit those unfortunate families. This money went directly to the affected families, with no deductions taken from it.

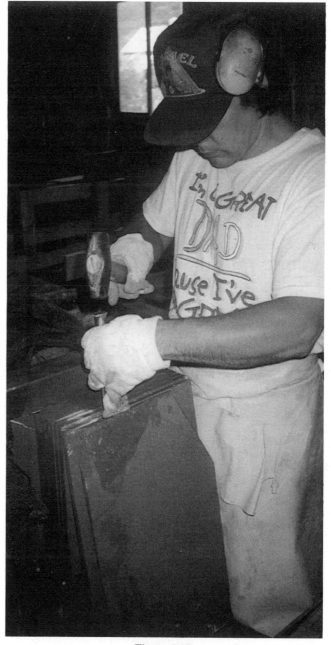

Figure 5.19:
Today's American slate splitters still use a hammer and chisel to split slate blocks into roofing shingles, as this splitter at Hilltop Slate Co., Granville, NY, demonstrates.
(Photo by the author)

Today's slate quarries are a world apart from the old mines that produced the slates for most of the older roofs in America. Modern quarries are much safer, and the workers are protected by insurance and other benefits. They now use diamond saws, forklifts, trucks, bulldozers, highlifts, and other modern equipment (Figure 5.18, 5.20, 5.23). The one operation still done by hand, however, is the splitting of the slate, still achieved by hammer and chisel (Figure 5.19, above).

Figure 5.20: ▲ Vermont Structural Slate Co. quarry pit north of Poultney, Vermont.

▼ A modern diamond saw cutting a block of slate at Vermont Structural Slate Co. in Fair Haven, VT.

[Photos by the author]

Figure 5.21: ▲ Mountains of waste slate like this one near Poultney, Vermont, where the author perches with his daughter, are a common sight in quarry regions. [Photo by Jeanine Jenkins] ▼ Finished roofing slates are waiting to be shipped out at the Evergreen Slate Company in Granville, NY (1993). [Photo by the author]

Figure 5.22:
▲ **Trim room, Evergreen Slate Company, Granville, NY (1993)**
▼ **Recycled slate ready to be shipped to restoration jobs, supplied by Camara Slate Co., Hampton, NY.**
(Photos by the author)

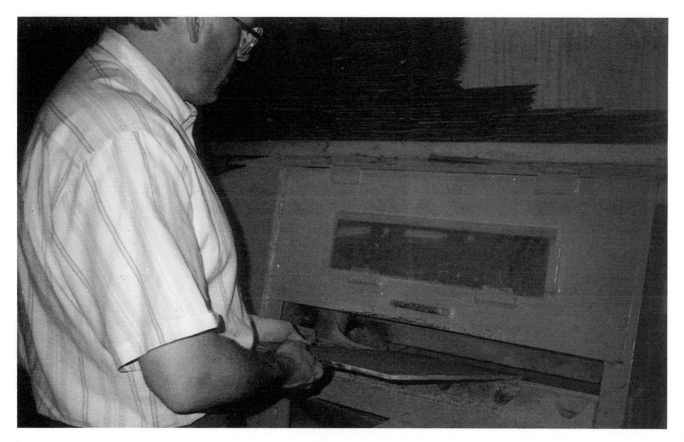

Figure 5.23:
▲ The guillotine-blade type of modern slate trimmer is demonstrated by
Clark Hicks, President of Evergreen Slate Co. in Granville, NY (1993). (Photo by the author)

Chapter 5 References

• Beers, F. W., (1969); <u>Atlas of Rutland County</u>, Charles E. Tuttle Publishing Co., Rutland, VT, pp. 111-114.
• Bowles, Oliver, (1934); <u>The Stone Industries</u>, 1st Ed., McGraw-Hill Book Co. Inc., New York and London, p. 278.
• Joslin, J., Fisbie, B. and Ruggles, F., (1875); <u>A History of the Town of Poultney, Vermont, From Its Settlement to the Year 1875</u>, Poultney: Journal Printing Office, pp. 181-187.
• Morrow, John A., (1970); <u>A Century of Hard Rock - The Story of Rising and Nelson Slate Company</u>, Grastorf Press, Granville, NY.
• *Ruggles Stone Machinery Catalog*; Ruggles Machine Co., Poultney, VT (Established 1828). This catalog provided the illustrations of the slate-working equipment in this chapter. Special thanks to Paul B. Boyce and the East Poultney Historical Society.
• Sharrow, Gregory (ed.), (1992); <u>Many Cultures, One People</u>; The Vermont Folk Life Center, Middlebury, VT 05753, pp. 232-235.
• Smith, H. P., and Rann, W. S. (ed.), (1886); <u>History of Rutland County, Vermont</u>; D. Mason and Co., Syracuse, NY, pp. 192-198.

Chapter Six
Pennsylvania Slate

"When the ribbon turns toward the mountain, that's the best slate."
Randy Rowe

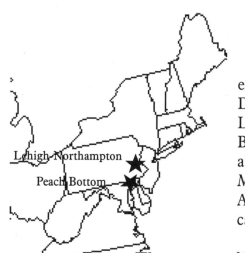

Pennsylvania slate quarries exist in two districts. The largest, most productive and well known of these is the Lehigh-Northampton district, which extends about thirty miles from Slateford in the north on the Delaware River, to beyond Slatington in the south on the Lehigh River, including the quarry towns of Bangor, Pen Argyl, Belfast, and Chapman, among others. The slate belt here is about five miles wide, lies on the south side of the Blue Mountain, and may approach 500 million years in age. Approximately 90% of the slate produced in Pennsylvania came from this district, which is still productive today.

It should be noted that this slate deposit extends into New Jersey, and slate quarries have been operated near Lafayette, in Sussex County, New Jersey by the Lafayette Slate Mining Corporation of Port Chester, NY. However, as the New Jersey productions were dwarfed by the Pennsylvania productions and are not in operation today, the New Jersey slate deposit will be considered in this book as equivalent to the Pennsylvania slate of the Lehigh-Northampton district.

The second slate district in Pennsylvania is the Peach Bottom district in York and Lancaster counties, which straddles the Pennsylvania - Maryland border along the Susquehanna River near Delta, PA (see map, next page). This slate deposit is 10 miles long and may be nearly a billion years old. Peach Bottom slate has a higher degree of metamorphism and is higher in silica, making it a harder, more durable slate than the average Lehigh-Northampton slate. In fact, Peach Bottom slate was once considered "the best slate in the world," but is no longer quarried. Peach Bottom slate will be discussed in greater detail in the next chapter.

Pennsylvania has long been the nation's leading producer of slate. Although at least ten states were producing slate as late as 1944, Pennsylvania's slate production averaged 40-50% of the nations output during the first four decades of this century. Today, however, slate production has markedly declined, not only in Pennsylvania, but also throughout the other slate producing regions of the US. This is due to the market competition of synthetic roofing materials, as well as the tremendous amount of work, and waste, involved in slate quarrying. Some estimates put the amount of waste rock in slate quarries at 70-90% before the invention of the wire saw in 1926, which subsequently enabled slate to be cut more efficiently (see Figure 6.3). In 1985, a full 65%

◀ **Slate block being hoisted to the surface, probably in a Pennsylvania mine, illustrating the degree to which quarry workers may be exposed to overhead hazards.**
Source: United States Department of the Interior, Bureau of Mines Report of Investigations 9009, 1986; p. 7.

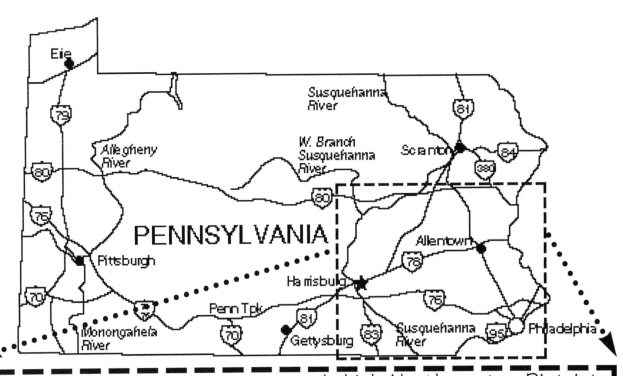

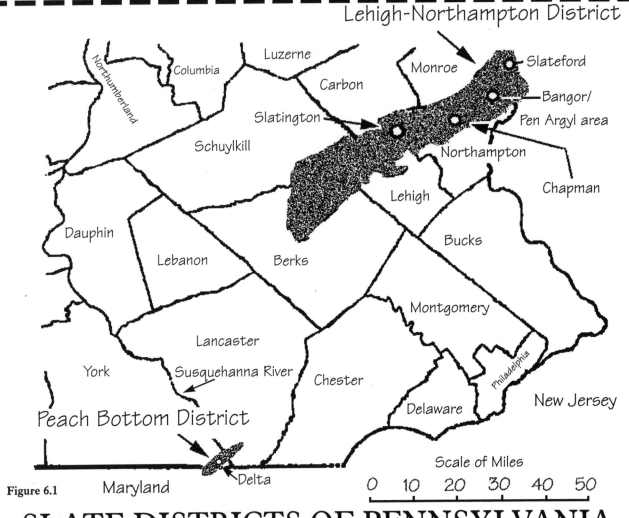

Lehigh-Northampton District

Slateford

Bangor/
Pen Argyl area

Slatington

Northampton

Chapman

Peach Bottom District

Delta

Figure 6.1

Maryland

New Jersey

Scale of Miles

0   10   20   30   40   50

SLATE DISTRICTS OF PENNSYLVANIA

[From: Slate in Pennsylvania, p. 6]

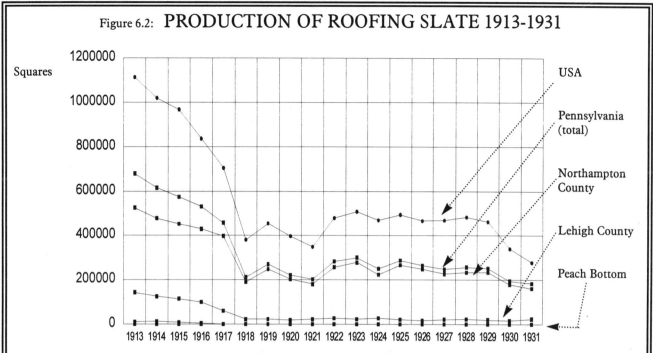

Figure 6.2: PRODUCTION OF ROOFING SLATE 1913-1931

The production of roofing slate in the United States dropped dramatically during WWI, never fully recovering. Pennsylvania has produced most of the roofing slate in the U.S., although relatively little Peach Bottom slate was ever produced. [From Slate in Pennsylvania, p. 112]

waste was *still* generally considered acceptable in the Lehigh-Northampton district. Due to the decline in business, many quarries have shut down and entire regions now lay idle.

Some slate regions adapted to the decline in demand for roof slate by producing slate granules for asphalt shingles. Nearly a million tons of roofing granules were produced in 1942, for example, and ground slate was also put to use as an industrial abrasive, as a filler in paints, asphalt mixtures, roofing mastic, oilcloth, and other products. Slate has other uses such as for billiard tables, blackboards and for structural purposes, and Pennsylvania has been the nation's leading producer of these slate products.

The slate of Pennsylvania is primarily black or dark gray. In the slate industry much of the PA slate is known as "blue-black," as it looks slightly bluish when first quarried. It ranges in durability from soft (S2) slate with as little as a 50 year lifespan, to an extremely hard (S1) slate with a lifespan that may approach four centuries or longer.

The vast majority of the Pennsylvania slate quarried for roofing is soft black slate with a life span on roofs ranging between 50 years and 125 years, originating in what is known as the "soft vein" region of Lehigh and Northampton Counties. This region includes Bangor, East Bangor, Pen Argyl, Danielsville, Slatington, and Slatedale. Only a relatively small amount of slate from the Keystone State originated at the famous Peach Bottom quarries (Figure 6.2). Other Pennsylvania slate comes from the "hard vein" belt, which occurs near Chapman and Belfast, and which falls somewhere between soft and hard on an overall slate scale. Then there is the "gray-bed" slate, which is not as durable as Peach Bottom, but may be considerably more durable than the Pennsylvania soft slate.

Unlike Maine slate which is commercially only one type of black slate, Pennsylvania's vast slate deposits produce many types of slate with many characteristics, although all are black (or close to it) and may not seem very different in appearance to the layperson. As mentioned in Chapter two, the main slate deposit in eastern PA, which produces primarily soft slate, is 2800

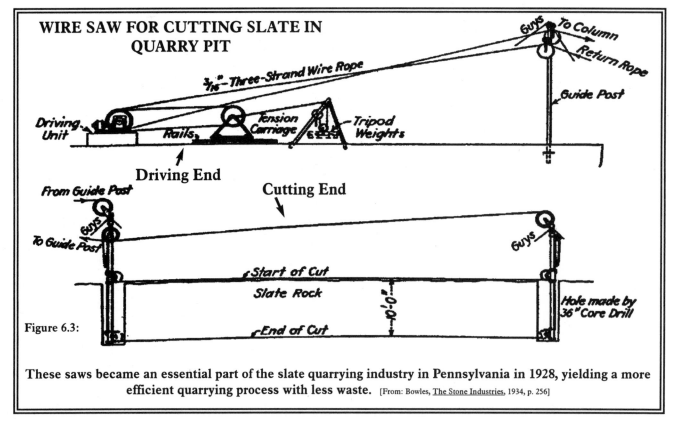

WIRE SAW FOR CUTTING SLATE IN QUARRY PIT

Figure 6.3:

These saws became an essential part of the slate quarrying industry in Pennsylvania in 1928, yielding a more efficient quarrying process with less waste. [From: Bowles, The Stone Industries, 1934, p. 256]

feet thick, two to five miles wide, and thirty miles long. This is such a massive enough deposit that considerable variation can occur throughout its range. In the Pen Argyl region, for example, the following slate runs appear in succession: the *Pennsylvania Run, United States Run, Diamond Run, Albion Run, Acme Run, and Phoenix Run,* each run bearing a slate of a different quality, although the difference may not be significant. The runs are separated by intervening beds 75 to 280 feet thick of unworkable slaty rock.

Each run is itself then separated into individual *beds.* The Albion Run, for example, consists of 12 beds combining to form a total run thickness of 184 feet. On a visit to the PA slate quarries in July of 1993, the author came upon an old list naming the veins of the Albion run. A quarry worker had evidently scrawled the list in pencil on a piece of cardboard and tacked to the wall of an old slate shanty, long since abandoned when the author stumbled upon it. After blowing the dust off the cardboard, I found a full 43 veins listed, each with a colorful name either pertaining to the qualities of the slate in that vein, or honoring a friend or family member (see Figure 6.4).

If you've read Chapter Two of this book, you understand how slate was formed and realize that the rock was originally clay which settled under water in layers over periods of millions of years. Because of the immense amount of time it took for those layers to form, they were subjected to different environmental factors over time, and today those layers of finished slate will show different characteristics as one digs deeper into them. As a rule, the deeper the slate, the older it is and the harder and more durable it is (providing that the slate deposit hadn't been heaved upside down by geological forces). This is one reason why Peach Bottom slate is more durable than the softer slates farther north in the PA slate belt. It is geologically much older.

In any case, the list of slate veins in the Albion Run near Pen Argyl, PA, in Figure 6.4 provides an excellent illustration of the variation that can be found as one digs deeper into a PA slate deposit. Figure 6.5 indicates more accurately how the layering of the slate may appear.

Figure 6.4:
Names of Veins in the Albion Run, Pen Argyl, PA

The top of the list is closer to the earth's surface. Names are printed as written. At the bottom of the list is a question mark to indicate that the men didn't know what they'd find as they dug deeper.

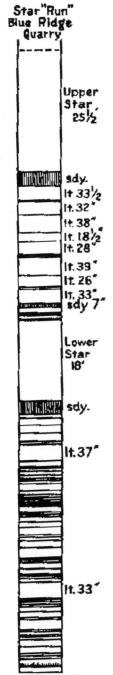

Figure 6.5:
Example of thickness measurements in beds in PA. Above diagram represents 71' of depth. ["lt." = light, "sdy" = sandy]
From <u>Slate in Pennsylvania</u>

When one understands that black slate from one quarry can have a variety of characteristics depending on the vein in which it was found, then it is easy to understand how difficult it may be to pinpoint the exact origin of a slate currently serving on the roof of a house. If you have a soft black slate on your building, then you can be pretty sure it is from eastern PA. Yet some black slate from PA is hard. And some slate from PA is gray and hard. It is the hardness of the slate that we are interested in, as a hard slate roof should be restored when old, but a soft slate roof must be replaced. Many thousands of roofs have PA slate on them, but exactly where in PA those slates came from may be difficult to say. There are some exceptions to this, however.

The first, of course, is that if you live near a slate quarry, you are likely to have *that* slate on your roof. If you live in Chapman, PA, you probably have a Chapman slate roof. Same for the Peach Bottom area (Delta, PA, for example) - you have a Peach Bottom roof. In Bangor, PA, you probably have a slate on your roof from the Bangor area. However, all of these slates, and many others from PA have been shipped all over the eastern seaboard of the United States during the last century. And if you don't live near a slate quarry and you think your roof is a PA slate roof, how do you identify it?

Peach Bottom roof slate is easily identifiable because it is black, remains *smooth* with age, and stays very hard. It also has a characteristic sheen to it that may not be evident unless you look at the back of the slate (assuming the front of the slate has been environmentally stained with age). See page 145.

Chapman slate is easily identifiable because it is black, somewhat hard, and has diagonal bands (ribbons) across the full width

of the slate that are evident from a distance - you can see them from the ground if the slates are on a roof (see Figure 6.7, page 106).

Bangor black is a uniform black, smooth slate with no *apparent* ribbons (although destructive carbon-bearing ribbons can be present, see Figure 2.8, page 26), and is fairly soft, but can still last a good century. With age these roof slates will appear to become flaky and will crumble, especially along those nearly invisible carbon ribbons. Many of the older Bangor slate roofs are reaching the end of their years at age 75- 90, and many have already worn out. Bangor slate was shipped with a "Genuine Bangor Slate" label (left and below) which some roofers left stuck to the slate when they installed the roof, so when the roof is removed 80 years later, the label is still there on the back of the slate making the slate readily identifiable.

Pen Argyl black slate may develop a light brownish surface color due to iron in the slate, and will also become flaky with age. In some cases, these slates can look remarkably similar to some old Vermont sea green slates to a layperson. People may point out old Pen Argyl slate roofs and mistakenly identify them as sea green roofs simply because of the "rusty" surface color that has developed over the years. When Pen Argyl slate gets old, however, it will be visibly deteriorating and individual slates will be seen falling apart on the roof. The pieces that fall to the ground will probably have soft edges, soft enough to crumble in your fingers. Old sea green slate, although changing to a brownish color, won't turn as soft so quickly.

Some of the soft black slates from this region exude a chalky substance over time and may eventually display a whitish hue, especially around the edges.

Slate from the "gray beds" of eastern PA seems to be among the most admired for durability, second, perhaps, only to the Peach Bottom slate. They are fairly rare, "olive green in color" according to one published source, but to me look dark gray (but not completely black), and contain little carbon. When carbon makes up the bulk of the "ribbon" in ribbon slate, it greatly reduces the life of a piece of roofing slate, which may be one reason why gray slates last so long in comparison to their black cousins. In fact, the gray-bed slates of PA are said to resemble the green slate of Vermont in chemical composition.

The first slate quarry in Pennsylvania's Lehigh-Northampton district opened in 1812 at Slateford in Northhampton County, according to one source. Another source claims the first quarry was opened in 1832 by

Samuel Taylor and actively operated from 1836 when James M. Porter joined him as partner. This occurred before railroad service existed into the area, and the slates were sent down the Delaware River on boats. This quarry only lasted about 30 years.

Another quarry near Slatford, the Snowden Quarry, opened in 1870 and operated until 1917, when it was deemed a non-essential war-time industry, and closed. During World War I many quarry workers went to the steel mills for war production, and many quarries closed and never reopened.

Slate was discovered in the Bangor area on a farm at what is now East Bangor, by Joseph Kellow in 1853. This quarry was operated from 1855 by fellows named Weidman and Derrick.

Bangor got its name from Robert M. Jones, a pioneer in the slate industry in the Bangor community, who took the name from his hometown of Bangor, in northern Wales. Jones opened the Old Bangor Quarry in 1863, and the town that sprung up around it took the same name. Once again the Welsh played the role of slate entrepreneurs, founding an industry in Pennsylvania from which entire communities emerged.

Today's PA slate quarries are huge, neatly cut holes that plunge vertically into the earth for hundreds of feet (see Figure 6.8). The abandoned quarries of this type have completely vertical cliff-like walls dropping as much as 650 feet to a pool of water at the bottom. These can provide quite a hazard for persons out for a stroll in the woods unaware of the one-way drop waiting for them should their foot slip at the edge of one of these abandoned pits. Local folks tell of deer inadvertently dropping off the edges of these cliffs on a one-way free-fall to a watery grave. Fortunately, accidents are not common despite the unfenced holes (knock on wood).

One advantage to deep pits is that the temperature remains a steady 55 degrees Fahrenheit at the bottom where the workers toil daily to wrestle out the slate slabs. This provides a relatively cool work site in the summer and a relatively warm one in the winter.

CONVERTING SLATE BLOCKS INTO SHINGLES

The slate in Pennsylvania is still split into roofing shingles by hand, and the blocks are kept wet with water during the splitting process to enhance the ease of splitting. The stone is said to split *"with much greater ease if the quarry sap is not allowed to evaporate,"* an opinion echoed by slate workers in Wales, who insist that slate blocks left out on the surface of the earth to dry will not split as readily as freshly quarried slate.

The splitting of slate into the right thickness for roofing shingles may not entirely be the responsibility of the splitter, but may involve the block cutter and the blacksmith. The block cutter can split the incoming block to the proper thickness to allow for a predetermined number of slates to be produced from one slab. Since the standard thickness of a roofing slate is 3/16 inch, the slab should be a factor of that thickness. In order for the block cutter to consistently split a block to the thickness desired, his chisel is conveniently made to the proper width and can be used as a measuring tool as well as a chisel. So the blacksmith is instructed to forge out a chisel that the block cutter can use to measure his cut. In this manner, the slate splitter (the guy who splits out the actual roofing slates) receives blocks that have already been split to a convenient size, which he then splits by eye through the center, then in half again, then again to end up with the finished pieces of 3/16" roofing slate (see Figure 6.14).

Naturally, this technique is not precise and some slates turn out to be a little thick or a

Figure 6.6:

▲ This is a typical sight on a PA slate roof that has been poorly maintained. The slates are disintegrating, but not being adequately replaced. Because this is a low-slope roof, it was probably repeatedly walked on, causing damage to the slate. There is little hope for this roof, although if it had been repaired with sound slates as the old ones broke, its life could have been stretched for years.

[Photo courtesy of U.S. Dept. of the Interior, National Park Service, Preservation Briefs #29, authored by Jeffrey S. Levine]

Figure 6.7:

▲ Chapman slate roof, good condition. Note "ribbons" or lines running across the face of the slate at an angle, which is a distinctive characteristic of Chapman slate. Chapman ribbons tend to be hard, and don't undermine the quality of the slate. Chapman slates, first quarried around 1860, were the most productive of Pennsylvania's "hard vein" slates. If cared for, these roofs can last well over 100 years. [Photo by the author]

Figure 6.8:
A block of slate is being hoisted out of a Lehigh-Northampton District slate quarry (near Pen Argyl, PA), appearing as a light colored rectangle in the top center of the picture. The white dot near the bottom of the photo, right of center, next to a black line, is a quarry worker, dwarfed in this 350' deep, vertical-walled hole. The photo on the next page shows the block emerging from the top of the pit. It will be taken into the mill where it will be cut and split to size, yielding Pennsylvania's characteristic black slate.

[Photo by the author]

Figure 6.9 Slate block being hoisted out of quarry en route to the mill where it will be split down into roof slates (as shown in Figures 6.13 and 6.14). (Photo by the author)

little thin. However, the finished slates shipped out from the slate yards in the old days were amazingly uniform when the shingles were split from a good quality slate - the better the quality, the easier it is to split a thin shingle. When stacked, one can routinely expect 50 slates per foot of row of slates that are standard 3/16" thickness. This figure (50 slates per foot of row) is close enough that a person who has a stack of standard antique American slates need only measure the total length of the rows, in feet, and multiply by 50 to get a close estimate of the number of slates (10 feet of slate shingles stacked tightly on edge will amount to 500 slates, for example).

And slate *should* be stacked *on edge* on boards or slats, and not piled in flat, vertical piles like dishes, as the weight of flat stacks can weaken and break some of the slates over time (See Figure 6.11). Roof slates are amazingly strong on edge, and can be thrown from a roof onto a concrete driveway without breaking, so long as they land *on edge*.

The final trimming of the slate shingles was commonly done using a foot powered trimmer with a straight, steel cutting blade about three feet long. The outer end of the blade was attached to a spring. Another type of trimmer had a rotary blade like an old-fashioned push-type lawn mower. Although these machines were once powered by foot or hand, today they're electrically powered (foot powered and electrical slate equipment are illustrated in Chapter 5).

Each slate shingle was laid against a steel bar on the trimming machine when it was trimmed, the bar being marked in inches so that the trimmer could trim the slate to the standard sizes, ranging from 6"x10" to 14" x 24." The person trimming the slate could easily estimate by eye the largest possible size for each untrimmed piece, and trim accordingly. Once trimmed, the slates were stacked in piles sorted by size, then handed to a person who had the job of punching

Figure 6.10: ROOFING SLATE SOLD IN PENNSYLVANIA - 1890-1941

The production of roof slate peaked in Pennsylvania around 1902, with the bulk of the production occurring between 1896 and 1915. Production continues to this day, and the deposits remaining in the ground have barely had their surface scratched. Many homes were built during that period of peak production, and many had (and still have) PA slate on their roofs. [Source: *Mineral Resources of the United States* and *Minerals Yearbook* (dates unknown)].

the nail holes in the slate (see Figure 6.15). The holes were punched into the back of the slate (the side facing down against the roof) due to the countersinking effect of the hole punch which enabled the nail heads to set into the slate. Two men on a good day can split and dress 12-14 squares of slate, if they have nice blocks to begin with.

CHAPMAN SLATE

It is worth mentioning the Chapman quarries, which are located (you guessed it) in Chapman, PA. The quarries are no longer in operation, but they were once considered the most productive producers of Pennsylvania "hard vein" slate which are still serving thousands of roofs today. This black slate has a distinctive appearance because it is adorned with many ribbons, or bands of various shades of black, that cross the face of the slate at an angle, giving the slate a striated appearance. These ribbons are fairly hard and do not disintegrate as readily as the carbon ribbons common to the "soft vein" slate of Pennsylvania. I have seen many hundred year old Chapman slate roofs still in pretty good shape, as these slates are one of the harder and more

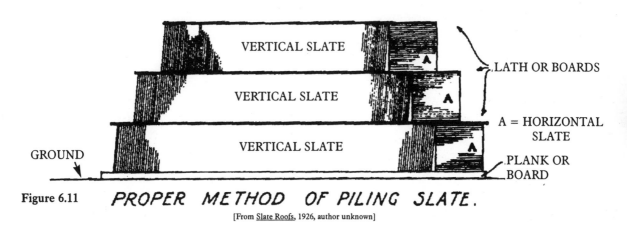

Figure 6.11   *PROPER METHOD OF PILING SLATE.*

[From <u>Slate Roofs</u>, 1926, author unknown]

durable of the Pennsylvania black slate (see Figures 6.7 and 6.12).

The Chapman Slate Company was one of the oldest producers of slate in Northampton County, with the Chapman Standard Quarry opening in 1860, and producing primarily roofing slate and slate slabs. Several other quarries operated in the Chapman vicinity. The "Chapman" Quarry consisted of two large holes separated by a wall of rock about 50 feet wide, the west hole being 450 feet long, 300 feet wide and 130 feet deep, while the east hole is about the same width but is about 1000 feet long, and 150 feet deep, rivaling the largest quarries in the Bangor and Pen Argyl districts in volume. This abandoned quarry pit still lies just outside Chapman, PA, today, and probably won't be going anywhere soon. There was still some quarrying reportedly taking place there in 1924, and in 1931 the Chapman Slate Company was still listed as a slate producer in Pennsylvania. Today, however, no quarrying is taking place there and Chapman roof slate cannot be bought new.

On the other hand, Chapman slates are one type of black Pennsylvania roof slate that are recyclable, and care should be given to remove these slates from roofs of buildings being demolished or re-roofed so the salvaged slate can be used to keep existing Chapman roofs in good repair. Otherwise, if recycled Chapman slates are not available, almost any black Pennsylvania slate can be used to repair a Chapman roof, so long as the slate is sound, although no slate will match the roof quite as well as a genuine Chapman slate.

It should be noted that the slate in Pennsylvania seems to become harder and more durable as we move farther south in the state. When we arrive at the border between Pennsylvania and Maryland we find a black slate so hard and durable that it is considered by many to be the best slate in the world, as we shall see in the next chapter.

Figure 6.12:
▲ Boro Hall of Chapman, PA. The town is not much bigger, but it is home to Chapman slate, although the quarries are no longer in production. The "hall" has a Chapman slate roof.
▶House near Chapman, PA, sided (and roofed) with Chapman slate.
(Photos by the author)

Figure 6.13: ▲ Slate blocks hoisted out of the Dally Slate Company quarry pit near Pen Argyl, PA, lay stacked and waiting in the modern slate mill. The edges have been sawed on a diamond saw. Some of these blocks will be used for structural slate, some for roofing.

Figure 6.14: ▼ The man below splits the slate into shingles the way it's always been done in America - with a hammer and chisel. [Photos by the author]

Figure 6.15: ▲ Nail holes are punched into the roof slate with an old-fashioned foot powered machine. Trimmed and finished Pennsylvania roof slate from the Lehigh-Northampton District (right, above) are stacked in a trim room waiting to be carried out to the slate yard.▼ (Photos by the author)

Figure 6.16:
Barn near Townville, PA, in poor repair. Main roof is Lehigh-Northampton (black PA) slate; designs are made of Vermont green slate. Inscription reads, "L. Jones - 1902."
(Photo by the author)

References - Chapter Six

• Behre, Charles E., (1933); Slate in Pennsylvania; Pennsylvania Geological Survey, Fourth Series, Bulletin M 16, pp. 19-21, 112, 173-187.
• Bowles, Oliver, (1934); The Stone Industries; 1st ed., McGraw-Hill Book Co., Inc., New York and London, pp. 229-289.
• History of Trinity Evangelical Lutheran Church, 1878-1928; pp. 10-11.
• Pennsylvania State University, School of Mineral Industries; Properties and New Uses of Pennsylvania Slate; The Pennsylvania State College Bulletin, Volume XLI, No. 30, July 18, 1947, pp 7-25, 134-137.
• (Author Unknown), Slate Roofs, (1926); available from Vermont Structural Slate Co., Fair Haven, Vermont 05743; phone: 802-265-4933/34 or 800-343-1900, fax: 802-265-3865. A version is also available from Hilltop Slate Co., PO Box 201, Middle Granville, NY 12849; phone: 518-642-2270/642-1453, fax: 518-642-1220.

PEACH BOTTOM SLATE REGION

PEACH BOTTOM SLATE, FIRST USED 1734, IS THE OLDEST IN AMERICA, THE FIRST COMMERCIAL CUT HAVING BEEN MADE 1785 BY WORKMEN WHO WERE PRIMARILY WELSH. AT THE LONDON CRYSTAL PALACE EXPOSITION, 1850, PEACH BOTTOM SLATE WAS JUDGED BEST IN THE WORLD.

MARYLAND HISTORICAL SOCIETY

Chapter Seven
Peach Bottom Slate
Pennsylvania/Maryland Slate District

Professor Agassiz (late 1800's) *"The Almighty might have made a more perfect fish than the trout, but he never did."*
Mr. Humphrey (quarryman) *"I say the same for the Peach Bottom slates."*

What intrigued me most about Peach Bottom slate was its mystery. In nearly three decades of working on slate roofs I had never seen a Peach Bottom slate (or so I thought) and hadn't a clue about what one looked like. One day someone phoned me and told me he had some roof slates for sale. This was a guy just five miles down the road who described the slates as "really nice, beautiful Peach Bottom slate." I drove to his country home to look at the slate and sure enough, there lay a pile of really nice slate in his front yard, neatly stacked, with scalloped bottoms. They were a light gray slate with a reddish cast to some of them, which I recognized as sea green slate from Vermont. I didn't buy them because the guy wanted too much money, but I wondered how he came to the conclusion they were Peach Bottom. Apparently he thought the color was peachy.

In my wanderings through the slate regions of Vermont, New York and Maine, I came upon old books about slate production in the United States, and Peach Bottom slate was mentioned as being of the highest quality, lasting hundreds of years. This was hard to believe as most of the slate from Pennsylvania is soft and relatively short-lived compared to the harder slate from Maine, Vermont, New York and Virginia. The notion that Pennsylvania, with all its soft slate beds, could also be the source of slate so hard and durable as to deserve such high regard by authors of pre-1900 slate books only added to the mystery. The mystery was compounded by the fact that Peach Bottom slate was no longer in production and could not be bought new on the roofing market.

Furthermore, modern maps of Pennsylvania show no town called Peach Bottom. However, I happened to have an *old* map of Pennsylvania and it did show a tiny little place named Peach Bottom just north of the Maryland border on the east bank of the Susquehanna River. So one day I couldn't stand it any longer and I set out for that spot on the map, a full eight hours drive from my home in western PA. I was determined to find the truth about Peach Bottom slate once and for all. I dragged a couple of my kids with me to keep me company, and one July morning we

Opposite: Road sign near Delta, PA, on the Maryland side of the PA/MD border. (Photo by the author)

took off. We drove all day until we reached a campground in Maryland just across the border from Pennsylvania, where we pitched a tent and got a little bit of sleep under the starry night sky. The loud serenade of summer's nocturnal creatures kept us awake for a while. My fifteen year old daughter's comment was, "I wish those birds would shut up!" "Those are tree frogs," I said, "Go to sleep."

Early the next morning we drove the back roads to the dot on the map named Peach Bottom. We found one or two houses and a boat marina on the east bank of the Susquehanna right across the river from the Peach Bottom nuclear power plant, which loomed on the opposite bank like a behemoth, its high powered electrical transmission wires emanating from it, octopus-like, in every direction. This didn't look like a place for a slate quarry, in my opinion, but I did notice that the name of the road going uphill past the marina was Slate Hill Road. "Let's follow this road," I said to the kids, like a bloodhound on a trail.

At the top of the hill we passed an old farm with slate roofed buildings (see Figure 7.2). We then came upon a small old house, also with a slate roof, where an old man was roto-tilling his garden out back. I pulled over and shut off the van. The old fellow turned off his roto-tiller and we met in his side yard.

"I'm looking for the old Peach Bottom slate quarries," I said, after introducing myself. "Do you know anything about them?"

"Why sure I do. What do you want to know?"

"Where are they?"

The fellow turned around and pointed off across the green fields of corn and hay to a patch of trees on top of the hill we had just driven by. "Back there in those trees you'll find the old slate piles." I pumped the friendly man for more information, but he didn't know much about the quarries themselves. He did have a pile of slate in his back yard matching the solid black slate on the roof of his house, so I bought a few. "This house was built in the 1700's," he said. "See that roof? That Peach Bottom slate is just as good as the day it was put on."

The slate itself is coal black and very hard, with a sheen to it that I had not seen on any other black slate. I passed roofs with these slates on the drive to Peach Bottom, seeing more of them as I got closer to my destination and wondering if they were, in fact, the elusive Peach Bottom slate. I had not seen any slate like this back in western PA (although once I learned to recognize them, I did begin to see them back home). The most striking characteristics of these slate roofs were the darkness of the black color and the shiny gleam.

The kids and I set out in the van for the wooded hill in the distance. We found a dirt trac-

FOULK JONES, President.
D. W. JONES, Treasurer.

MARSHALL F. JONES, Secretary.
JOS. S. WHITEFORD.

Slate Hill, Pa. _____ 190

ℳ _____

Bought ...of **Foulk Jones & Sons,**
Miners and Manufacturers of the Genuine Old

❧ **PEACH BOTTOM ROOFING SLATE.** ❧

TERMS :

Bills Payable only to the firm, or D. W. Jones, Treasurer.

Old payment receipt for Peach Bottom slate.

Figure 7.1:
This manse, built in 1954 next to the Chestnut Level Presbyterian Church, was probably one of the last houses to get a new Peach Bottom roof, although the slates may have been recycled. The church, built in 1765, also has a Peach Bottom roof.

[Photo by the author]

tor road leading up to the woods, and there we pulled off and set out on foot. The air was hot and muggy and puddles littered the lane, which was all but washed out from recent heavy rains. We were sweating by the time we got to the top of the hill, probably a half mile walk between corn fields and hay fields thick with huge butterflies. We stopped at a giant electrical transmission tower on a bluff overlooking the beautiful Susquehanna River. The nuclear power plant was now directly opposite us, connected to the tower by electrical lines hanging across the water, cracking and buzzing menacingly over our heads. My hair all but stood on end. From our vantage point I could see a mountain of slate rubble lurking in the trees across the corn field, so the kids and I wove through the rows of corn to get a closer look. When we reached the edge of the woods I plunged into a forest of poison ivy, wading through it to get to the slate pile. The kids were wearing shorts so I cautioned them to stay out of the woods and go back to the van and wait for me there.

I climbed the huge pile of scrap slate like a mountain goat and stood on top of it like some sort of conquering explorer, peering with some trepidation down the cliff-like bank of the river, on the brink of which I now perched. I had found the source of Peach Bottom slate! Or had I? Where was the pit? All old slate quarries had huge holes in the ground, usually filled with water, as well as huge piles of scrap slate. There was no visible hole here, although there had to be one somewhere nearby. I was reluctant to explore further through these dense woods, however, as the ground was covered with one of the thickest and most luxuriant growths of poison ivy I had ever seen, and my kids were now left unattended a half mile away along a rural roadside. The sweltering heat added to my desire to return to the van, so I picked up a heavy chunk of slate and carried it back to the road where I found the kids patiently waiting. "Let's stop at that

Figure 7.2: The Peach Bottom slate on this Weikel farm house dates from 1806, although the rear extension dates to 1796. The roof slates are still in very good condition.

[Photo by the author]

farm we passed and see if we can photograph their slate roofs."

The middle-aged farmer, Allen Weikel, at that time a write-in candidate for Governor of PA, explained that much of the slate was actually quarried on the other side of the river. He was a very friendly fellow who took a good hour out of his busy day to make phone calls trying to locate someone who knew about the history of the quarries. Soon we hit paydirt. We had Roger Faill of the PA Dept of Environmental Resources Geologic Mapping Division on the phone, and his advice was accurate and to the point: "The museum in Delta. That's where you should go. I've never been there, but I hear that they have some good information on Peach Bottom's history." So off we went to Delta, PA, right on the Maryland border on the west side of the Susquehanna, the heart of the Peach Bottom slate region.

By this time it was after noon and we were all getting hungry. As soon as we reached the quaint little town of Delta we stopped at the diner on the main street. I ordered their lunch special, and while waiting for my soup I asked the waitress if she knew anything about the old Peach Bottom slate quarries. "I know just the person you should talk to," she said. "Hold on a minute, let me see if he's still here." She disappeared into another room and soon returned with a wiry, elderly man named Harold Beucker, who pulled up a chair and began to fill me in on what he knew.

"I was the last person to haul Peach Bottom slate out of here," he said. "That was in 1956 or 57."

"Roof slate?"

"Yes, and I hauled it into the Pittsburgh area. Fox Chapel, if I remember right. Best slate ever produced. Best slate in the world."

"Why did they stop quarrying it?"

"Not economical. Too much waste, too hard to work, to split. Some of it went underwater when the river was dammed. After they quit making roof slate they continued to make slate granules for a few years. They used the granules for roofing, like what you see on shingle roofs."

"Where can I find out more about the history of the quarries?"

"The person you want to talk to is Ruth Ann Robinson. She lives just down the street. First house past the Welsh church. The red brick one. She's curator of the town museum. Museum's only open on weekends though. You won't be able to see it today."

After we finished eating, the kids and I set out for the Robinson house. No wonder I couldn't find an old quarry pit by the river, I mused as we walked, it was probably under water! The first thing I noticed when we reached the Robinson house, besides the Peach Bottom slate roofs on the church and the house, was a Welsh flag bumper sticker on the car parked in her driveway. Turns out that Ruth and her husband Don had both visited the slate areas of Wales. Don was making wooden slate-splitting hammers, like the ones used in Wales, in the workshop behind his house. Once again, a kind and friendly person, Ruth Ann, took time out of her busy day to accommodate total strangers with absolutely no advance notice. She opened up the Old Line Museum and dug around for material to show us, and some of the historical information about Peach Bottom slate in this book is attributable to her.

One important fact that became clear to me during my investigation was this: Peach Bottom slate is from Pennsylvania *and* Maryland. The slate deposit straddles the state line and it is a resource shared by the two states. In fact, the slate was mined in Maryland longer than it was mined in Pennsylvania. It was all shipped out of Delta, PA, though, and therefore became known as a Pennsylvania slate. So now I'm trying to set the record straight. Peach Bottom slate is a Pennsylvania *and* Maryland slate. Marylanders are rightfully proud of their slate, but haven't been given credit where credit is due. In fact, I had to go to Maryland to photograph the highway sign, paid for by Maryland tax dollars, shown at the beginning of the chapter.

The Peach Bottom slate deposit extends about ten miles, running northeast to southwest in three parallel belts 75 to 120 feet thick, and crossing both the Susquehanna River and the PA/MD border near where the Susquehanna flows out of Pennsylvania (see Figure 6.1, page 100). The slate deposit, locally known as a slate "ridge," extends from York County, PA, about 3 miles into Cardiff Township, Harford County, MD. Its total depth is estimated at 1000 feet, and its width is said to be one-fifth to one-half mile, although it seldom exceeds 200 feet in a quarry. About one and a half miles of the slate lay underneath the river until the construction of the Conowingo Dam submerged more of it. The quarries were located near Delta, PA, Peach Bottom, PA, and Cardiff, MD.

Slate splitters had some problems getting Peach Bottom slate to split to their satisfaction, and eighty-eight percent of the slate mined in the Peach Bottom district became waste.

Peach Bottom slate was exhibited in the Crystal Palace Exposition in England in 1850 by quarryman Roland Perry and was deemed the best roofing slate then known. It is perhaps this award more than any other that gained Peach Bottom slate its fame. Ironically, and despite its durability, Peach Bottom slate is no longer quarried. It's just too hard to produce finished slate products with this material at a price the market will bear. Perhaps this will change in the future and people will once again value durability over convenience and renew the markets for such fine slate products. Time will tell.

But don't hold your breath waiting. One recent summer a beautiful and rare Peach

Bottom roof on an old historic Inn in a small local town was damaged by a windstorm. I heard through the grapevine that contractors had been hired to tear the roof off and replace it with fiberglass shingles. I knew this would cost the Inn owner many times the amount it would cost to just repair the roof so I contacted the Inn manager and informed her that the Inn would be losing one of the best roofs in the world. She passed the word on to the owner, and it turned out that he didn't care, because his insurance was paying for the replacement roof. I then contacted the roofing contractor and offered to remove the slate, free of charge, so I could salvage them. I explained that these slates would still last hundreds of years, that they're no longer quarried, they're rare, and he would save a lot of work if someone removed the slate for him, etc. I told him that if he didn't like the idea of me working on his roof job, then he could take the slate off himself and I would buy them from him. At the very least, I said, I could be there when the slate was ripped off and gather them up before they went into a dump truck to be hauled to a landfill. The contractor said he'd get back to me. A couple of weeks later I drove by the Inn to see the last of the Peach Bottom slate being smashed into a dumpster, while the last section of roof was having fiberglass shingles *stapled* to it. When I stopped and gave the contractor a piece of my mind, he didn't seem to understand what I was saying. Perhaps I was speaking the wrong language, but then I don't speak *Neanderthal* all that well.

I'm mentioning this incident as a warning to homeowners. *Contractors may not care about your roof!* Believe me, there are many contractors out there waiting like vultures to tear off your good slate roof, destroy the slate, and staple glorified tar paper on

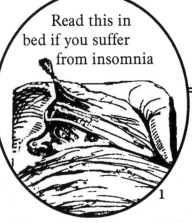

Read this in bed if you suffer from insomnia

Figure 7.3
Analyses of Peach Bottom Slates
From Five Different Quarries (%)

	1	2	3	4	5
SiO_2	55.88	58.37	60.32	60.22	44.15
TiO_2	1.27	tr.	n.d.	n.d.	n.d.
Al_2O_3	21.85	21.99	23.10	19.56	30.84
FeO (Fe_2O_3)	9.03	10.66	7.05	5.24	14.87
CaO	0.16	0.30	-----	3.87	0.48
MgO	1.50	1.20	0.87	2.30	0.27
K_2O	3.64	1.93	3.83	2.90	4.36
Na_2O	4.46	-----	0.49	2.15	0.51
CO_2	n.d.	0.39	n.d.	n.d.	n.d.
FeS_2	0.05	n.d.	0.09	n.d.	n.d.
C	1.79	0.93	n.d.	n.d.	n.d.
SO_3	.02	n.d.	n.d.	n.d.	n.d.
S	n.d.	.11	n.d.	.30	n.d.
MnO	.58	tr.	n.d.	n.d.	n.d.

[From: Slate in Pennsylvania, p. 375]

1901 advertisement

instead. They care about the money they're going to make, not about tradition, craftsmanship, durability, history and other things of character. Once your permanent roof is torn off and destroyed and you have it replaced with an inferior roofing product *guaranteed to fail* in 20 years or so, you are locked into a cycle of roof replacement that will continue to put money into the pockets of the contractors forever.

Furthermore, when insurance is paying for something, many contractors will push the homeowner to go for the most expensive option, no matter how wasteful. In other words, *if you don't protect your own slate roof, no one else will do it for you!* In this case I could say that the home owner deserved to lose the roof. But I saw this roof as an historic part of the community, and the building as one that may last for generations and have many future owners. It is disappointing when the owner of an historic building is not considerate of the community and not concerned about wasting good resources. Although this is happening every day in the United States, it is not legal in Europe to tear off a perfectly good, traditional stone roof and staple a tar paper product on instead. It is considered a breach of community standards and therefore prohibited. The entire community has a stake in the appearance of its buildings, and although this may be considered an unacceptable intrusion on personal liberties here in the USA, in Europe many quaint towns and villages continue to display the elegant charm they have demonstrated for centuries. Why? Because their building codes display a higher set of community standards which take into account such things as tradition and aesthetics. I fear that by the time we in the United States have evolved to that level of standards, there will be few, if any, slate roofs left.

Peach Bottom slate was more expensive than the other slates available at the turn of the century (1800-1900), so it had a tendency to be used on more expensive buildings and homes. It is certainly found on common buildings *near the quarry district*, but farther away it was sought after by people who had the extra money to spend for the very best material available. Therefore, many ornate homes, churches and government buildings had Peach Bottom slate on them, and today if you have an older fancy home or building with a hard, black slate on it, it may be Peach Bottom, unless the home is located near the Monson, Maine, or Buckingham, Virginia, slate districts (as these slate are also black and hard). For example, some buildings that had Peach Bottom slate on them by 1898 were the Pennsylvania State Capitol (roofed in 1820), Carnegie Library, Johns Hopkins University and Hospital buildings, Notre Dame Academy, the Biltmore Estate of George Vanderbilt, and the residence of William B. Astor on Fifth Avenue in New York City (see page 145 for a color photo of a Peach Bottom roof on the Buhl Mansion).

HISTORY

Peach Bottom slate was first discovered in 1734 by Welsh brothers who had settled in the area a few years prior. They used the slate to cover a roof, and the slates reportedly remained intact for 200 years until 1932 when the building burned. The roof was still in perfect condition. One history book tells the story like this:

> "Two brothers, William and James Reese, natives of Wales, took up land in 1725 under patent of the English provincial Government, in what is now known as York County, PA, and during the year 1734, when excavating for foundations for farm buildings, they discovered the slate rock from which they split the pieces necessary for roofing their buildings. This was the first use of slate in America." [From Peach Bottom Roofing Slate, 1898, p.5]

Those same pieces of slate were still in use in 1898 when the above historical account was published, on their *seventh* building, apparently having outlasted six others. *"They show no disintegration or decay* [after 163 years] *and still retain the strong, brilliant, blue-black color for which this slate has become famous by the name of Peach Bottom."* That seventh building was reported to be a hog pen located near Delta, PA, and still standing in 1930. It is reported in The Stone Industries (1934) that a sample of this slate was *"rescued from this lowly use and is now on exhibit at the United States Bureau of Mines in Washington D.C. After nearly 200 years of service it shows no evidence of deterioration."* Take a look at the photo of the one inch thick, 250 year old gravestone illustrated in Figure 2.10 (page 28) for an example of the durability of this slate.

One of the first slate roofs made of Peach Bottom slate was laid by Joseph Hewes, a signer

PEACH BOTTOM ROOFING SLATE.

We wish to emphasize facts already stated in connection with the Genuine Peach Bottom Slate.

1st. It is conceded to be the Best Slate in the World.

2nd. It is the *Oldest Slate in America.*

3rd. It has been *164 years in actual use.* Nearly *100 years longer* than *any other.*

4th. It *does not fade,* but *grows darker* with age.

5th. It is the *toughest* and *strongest* slate *in the World.*

6th. It *does not break* on the roof nor is it affected in any way by any extreme of temperature, hence no repairs needed.

7th. It is the most economical slate to use.

8th. It is the *only Genuine Peach Bottom Slate.*

9th. It is *manufactured only at Delta, York Co., Pa.*

We do not let our stock run low and are always ready to fill large orders for immediate shipment.

How To Be Sure To Get It.

By specifying it as "the Genuine Peach Bottom Slate."—See page 16.

By *requiring our certificate with each shipment.*

By *informing us* that *you have specified* it ; we will *watch it* up for you.

By *seeing* that the *Bill of Lading* reads from either *Delta* or *South Delta, Pennsylvania,* via Y. S. R. R., or *Cardiff* or *Cambria, Maryland,* via the B. & L. R. R.

Write to us for prices and information.

PEACH BOTTOM SLATE PRODUCERS ASSOCIATION,

Delta, York County, Pennsylvania.

Figure 7.5: The above advertisement is from a publication dated December, 1898:
"The History and Characteristics of the Peach Bottom Roofing Slate"

Figure 7.6
This octagonal roof (left) is being hoisted by crane onto the Caparosa house tower addition in Butler, PA. The main house already has a historic Peach Bottom slate roof (closeup below). The tower will be slated with Buckingham slate from Arvonia, Virginia, which today most closely matches Peach Bottom slate both in appearance and durability. Photo at right shows the Buckingham slate being installed on the tower by Barry Smith, who is wearing a safety harness attached to the top of the tower by a safety rope. (See also center color photos for a picture of a Peach Bottom roof.)

(Photo at left by Kevin Caparosa, other two photos by author)

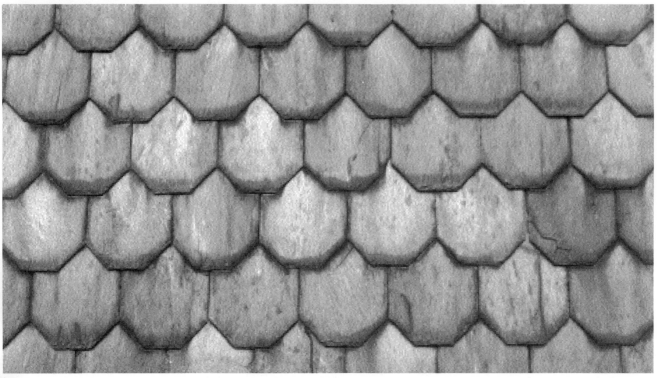

of the Declaration of Independence. The first commercial quarrying began around 1785 by a fellow named William Docker, but quarrying didn't get into full swing until the Welsh stepped into the scene in large numbers around 1845. In 1848 about 30 Welsh people arrived in Philadelphia en route to Peach Bottom to become slate quarriers, splitters, blacksmiths, and stone carvers. The Welsh introduced quarrying methods brought over from the quarries in Gwynedd, North Wales (see Chapter 4).

By 1858, 18 quarries were reported in operation west of the river and 11 east of it. At one time, thirty Peach Bottom quarries or prospects were listed west of the Susquehanna River and twelve east of the river (including the remains of one I "discovered" with my kids). By 1929, only three Peach Bottom roofing slate operations existed in Maryland, and none in Pennsylvania, and a year later only two remained in Maryland. In 1997, there were no slate quarries producing Peach Bottom roofing slate.

To illustrate the strength of Peach Bottom slate, a compression test was conducted in 1893 indicating that the slate would crush at 385.6 tons per square foot of pressure applied parallel to natural cleavage, and 758.4 tons per square foot applied perpendicular to natural cleavage. Another set of tests (1892-94) showed an average crushing strength of 11,260 pounds per square inch, a figure typical of a good quality slate.

Because Peach Bottom slate had gained fame due to its award at the Crystal Palace Exposition in England in 1850, other quarries began to falsely refer to their slate as Peach Bottom as well. These imitations went by the names of Slatington Peach Bottom, Lehigh Valley Peach Bottom, and Peach Bottom of Lehigh Valley, and were decried by producers of "genuine" Peach Bottom slate as "cheap imitations named to deceive the public," which they probably were (see Figure 7.5).

Today's lack of existing Peach Bottom quarries dictates the need to carefully salvage and reclaim Peach Bottom slates from buildings that are scheduled for demolition, or even scheduled for re-roofing. However, there is still one type of slate quarried in the USA today that closely resembles Peach Bottom both in appearance and durability, and which can be used with satisfactory success as a Peach Bottom substitute (see Figure 7.6). This is the "Buckingham" slate of Virginia, which we discuss in the next chapter.

References - Chapter 7
• Behre, Charles E., (1933); Slate in Pennsylvania; Pennsylvania Geological Survey, Fourth Series, Bulletin M 16; pp.359-390.
• Eisenberg, H. O., "The Story of Slate"; Slate Centennial; pp.13-16. (Date unknown).
• Faill, Roger T., and Sevon, W. D. (Eds.), (1994); "Guidebook for the 59th Annual Field Conference of Pennsylvania Geologists," specifically an article by Berkheiser, S.W., entitled "Some Commercial Aspects of the Peach Bottom Slate: The Problem of Being Too Good"; pp 143-145.
• Norris, John C., (1898); History and Characteristics of the Peach Bottom Roofing Slate; pp 4-16.

Advertisement from an 1877-78 Richmond city directory.

Chapter Eight
Virginia and Georgia

Virginia slate (Arvonia-Buckingham slate belt) is black, very hard and durable, is guaranteed by a manufacturer to last 150 years on a roof, and will probably last much longer. It has a glistening sheen indicative of a high level of quartz which sparkles in the sunlight when the slate is new. This slate is the closest thing to Peach Bottom slate that the author is aware of, and it is still quarried today at the little town of Arvonia, Virginia, deep in the rich heartland of the state.

This "lustrous" slate is said to be "blue-black," "deep black," or "oxford gray," however, for our purposes "black" is close enough. It is said to have the lowest average absorption (less than .02 per cent) of any American slate when submerged in water, which is a very good indication of longevity. It is also claimed to have equal strength both with and across the grain, which is unusual for slate. A manufacturer also states that this slate shows the least deterioration of any other slate in corrosive tests. Add to this the fact that the slate does not change color with age, and you have, all in all, one of the best slates on the market today, rivaling Peach Bottom slate (which is not on the market), Monson slate (also not on the market, although the Glendyne Quarry in Quebec now offers a substitute), and the best green, purple and red slate of Vermont/New York. In other words, this slate should last the life of the building it is nailed to. And if the building wears out before the slate does, the slate can be removed and used on another building. In some cases, as is done in Europe, the slate may need to be removed from an old building, the old roof boards removed and replaced, then the same slate renailed to the new roof boards on the same building with new nails. In this way a hard slate roof can be preserved indefinitely.

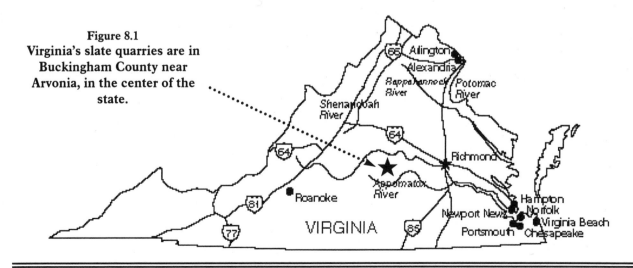

**Figure 8.1
Virginia's slate quarries are in Buckingham County near Arvonia, in the center of the state.**

Figure 8.2:
(Left) Two of the author's kids examining blocks soon to be split into black Virginia roofing slates, at the Buckingham-Virginia slate quarry in Arvonia. (Right) This Welsh gravestone in Arvonia dated 1886, is made from a thin piece of local slate and shows no deterioration after 110 years in the weather. The spots on the stone are raindrops. (Photos by the author)

The slate bed at Arvonia is 600 to 800 feet thick, 10 to 15 miles long, and 1 to 1 1/2 miles wide. It extends across parts of Buckingham and Fluvanna counties, and crosses the James River near the town of Bremo Bluff. Although often referred to as "Buckingham" slate in today's slate industry, it is known geologically as "Arvonia" slate on state maps. The slate lies in three distinct veins, the first having such a fine grain that it can hardly be detected, the next having a coarser grain, and the third vein remains undeveloped.

The slate has an overburden of shale (sitting over the slate bed) that is 40 to 60 feet thick, according to David Crummette, President of Buckingham-Virginia Slate Corporation (1994). This overburden constitutes the only waste derived today from that quarry's open-pit operations as the slate refuse is crushed for road aggregate.

Although many roof slates from this area are made from single beds, others show ribbons ("ribbing") indicating they were cleaved through a succession of beds, giving the slate an attractive striated appearance, but not affecting durability in this type of slate. Some thin beds only a few inches thick contain fossils of crinoids, brachiopods, and trilobites considered to be of Ordovician age (500 million years old). Other fossils including clams have been found. The deepest quarry in the district was reported to be over 350 feet deep and depths of over 200 feet are not uncommon. One of the largest single slabs of slate from the Arvonia-Buckingham slate belt trimmed to seven and a half feet by eleven and a half feet in width and length, and was able to be

Figure 8.3: Buckingham slate quarry, Arvonia, Virginia, after a heavy 1994 rain. (Photo by the author)

split to only one and a half inches in thickness!

Slate deposits are also found in the Esmont area in Albemarle County, in the Warrenton area of Fauquier and Culpeper Counties, and in the Quantico slate belt in Stafford and Spotsylvania Counties, but these are inactive.

Here again, as we have seen throughout the other major slate producing regions of the United States, the Welsh started the slate quarries. The name "Arvonia" for example, comes from "Arvon" which is a derivative of "Caernarvon," the name of the county (and city) in North Wales from which many a slate worker emigrated. A "caern" is a "fort" in Welsh, and the name *Caernarvon* was appropriately shortened to just *Arvon* when the Welsh settled in Virginia, eventually to again be altered to its final form, *"Arvonia,"* for reasons unknown. A principal town in Caernarvon, Wales, is Bangor, and *that* name was given to both Bangor, Maine, and Bangor, Pennsylvania, by Welsh immigrants.

Early reports of the use of slate in Virginia are sketchy, and by all accounts the first slate roofs came to Virginia from England. In 1662, the Virginia General Assembly authorized the building of 32 houses at "James City" and specified slate roofs. At that time, however, slate was not being procured from Virginia's own deposits. Likewise, in 1706, slate was again specified for the roof of the Governor's house in Williamsburg, authorizing its purchase from England. During the mid 1700's, the *Virginia Gazette* published notices of shipments of slate from Germany, Ireland and Algiers.

By 1755, slate roofs were being described in print, along with lead, shingle and tile roofs, on Virginia houses, and although Virginia's slate deposits were known then and were probably worked to some extent, it is not clear what percentage, if any, of these roofs were made of Virginia slate. In 1794, a contract was awarded to a roofing firm for the purpose of slating the

Figure 8.4:

▲ **David Crummette, President of Buckingham Virginia Slate Corporation, Arvonia, Virginia (1995), demonstrates the use of a modern electrically powered slate trimming machine with a "guillotine" style blade. Untrimmed pieces sit to his left, foreground. (Right) The trimmed slate.** (Photos by the author)

roof of Virginia's Capitol with slate from a Buckingham quarry, stipulating that the job had to be completed by 1796. Again, for reasons that are not clear today, that contract was never honored, and the Capitol was instead roofed with wooden shingles.

It is known that at least one quarry was in operation in Buckingham County by 1812. Unfortunately, just about that time, Thomas Jefferson, noted for (among other things) his architectural genius, decided that "slate is a bad covering because [of its] constantly getting out of repair." He preferred roofs covered with tin instead, approving of a roof design that only rose vertically a half inch in each foot of horizontal run, which is practically flat by today's standards. So, for a while, major roof contracts were let out to tin roofers rather than slate roofers, and Jefferson advised *"to cover with tin instead of shingles. It is the lightest and most durable cover in the world. We know that it will last 100 years, and how much more we do not know."*

Well, Jefferson's advice was wrong and, unfortunately, disastrous. The tin roofs lasted only ten years, and when they were replaced, the owners specified slate, one person noting in his diary that he had *"commenced taking off roof of the house to be replaced by a new one to get rid of the evils of flat roofing and spouts and gutters, or in other words to supercede the Jeffersonian by the common sense plan."*

In 1835 the tide was turning for Virginia slate, and geologist W. B. Rogers reported to the Virginia legislature that *"in texture, density, and capacity of resisting atmospheric agents [Buckingham*

Figure 8.5
Century old advertisement for slate exporting services.

County slate] can scarcely be excelled by a similar material in any part of the world."

By 1851, the slate was receiving "honorable mention" at the London World's Fair in England, followed in 1876 by a gold medal at the Philadelphia Exposition, and yet another medal in 1893 at the World's Columbian Exposition.

Perhaps the greatest factor involved in getting Virginia's slate industry moving was the Welsh influence. In 1867, as Virginia still smoldered from the Civil War, Welsh immigrants from North Wales began to move into central Virginia in search of opportunity in the slate quarries. Some of the Welsh people came down from Vermont where they had settled for the same purposes. By 1881, one Welsh quarry employed 100 mostly Welsh workers, when a week's shipment of roofing slate may have equaled a hundred tons in weight, traveling as far as San Francisco by rail. The advent of the railroad into the slate district in 1885 opened the final door to the markets, and the slate industry flourished, booming between 1900 and 1910, approximately the same time slate production peaked in Pennsylvania.

Once again, World War One had a bad impact on the slate industry, and of eight Virginia slate quarrying companies existing before the war, only four continued after. By 1928 there were only three companies in operation, but with the introduction of earth-moving equipment production actually rose to new highs. Still, the Great Depression caused quarry operations to virtually cease in the early 1930's, when, in 1932 slate production was as low as it had been in 50 years. By 1940 the three companies had recovered sufficiently to employ about 420 workers, but then along came World War Two, grinding operations to a halt, after which operations resumed at one third the pre-war level.

By 1962, however, 100,000 tons of finished slate products were being shipped from Buckingham County, Virginia, to all parts of the United States.

Today, although the immense deposits of slate in Virginia have hardly been touched, only two companies quarry Virginia slate for roofing. One is the Buckingham-Virginia Slate Corporation in Arvonia, which, at times, has a two year backlog of orders for roofing slate. I once ordered slate from them for a tower roof (page 124), and I only had to wait a few weeks for the order to be filled. This was a small order, however, and larger orders may require plenty of advance notice. To meet the demand for Virginia's high quality slate, another quarry has opened in recent years, the Virginia Slate Company (see page 218).

GEORGIA

Georgia slate deposits suitable for roofing slates are found in the northwestern corner of the state near the Cohutta, Silicoa, Pine Log and Dug Down Mountains in the southern Appalachian range. The quarries were localized in two areas: at Rockmart, in Polk County, and at Fairmount in Bartow County. Both quarry areas initially produced roofing slate, although both subsequently manufactured only crushed slate products (aggregate), until eventually all production died off in the 1970's (Fairmount) and 1980's (Rockmart). Slate deposits have also been reported in Murray, Gordon and Fannin Counties. Lindsay (1974), reports that the principal slate producing states in America in 1897 included Georgia, and Lindsay's *History of the North Wales Slate Industry* strongly implies that Georgian slate was exported to Europe at that time.

Figure 8.6

ROCKMART

Georgia's two quarry regions produce a distinctly different type of slate. The Rockmart slate, debatably of Paleozoic age (between 230 million and 600 million years old), is black, or "blue-black" (actually gray) very similar to the slate of eastern Pennsylvania's Lehigh-Northampton "hard vein" district. Some Rockmart slate remains quite hard and durable a century later, and appears similar to Peach Bottom slate, while some seems to be softer, becoming rather soft after a century. Apparently, the different grades of Rockmart slate depended on their original location within the actual quarries, the harder slate coming from one pit or tunnel, and the softer from another.

Much of the slate quarried has been ground up for aggregate and used as a concrete filler, asphalt shingle coating, etc., due to the lack of a smooth textured, finely grained slate suitable for splitting into shingles. As a result, there is not an abundance of Georgia slate roofs in the Rockmart area at the end of the 20th century, and many of those that still exist tend to be, unfortunately, in poor repair.

According to the Geological Survey of Georgia (Bulletin 27, 1912), Rockmart slate is a "very bluish-gray color and of slightly rough and lusterless surface." The survey adds, "This is a mica slate of the fading series [will change color somewhat with exposure to the weather], although some obtained from a tunnel at one of the quarries is reported to have kept its color for many years (see Figure 8.8, page 134). Its fissibility [ability to split] is fair."

Once again, the development of the slate industry in the Rockmart area was attributed to the Welsh (pronounced "Welch" by the locals, with a hard "ch"), although the first quarry was opened around 1850 by a Mr. Blance, who discovered slate on his own plantation in 1849. Mr.

Figure 8.7:

▲ The Rockmart Presbyterian Church, built of slate, displays the gray Rockmart slate on the bell tower. (Right) Gravestone at the old Methodist Church cemetery in Van Wert marking the grave of a soldier who died in 1861. The slate is still in very good condition, although many of the old slate gravestones in this cemetery have eroded beyond recognition, proving that Rockmart slate is both hard and soft, depending on the quarry site.

▼ Typical Rockmart slate roof.

(Photos by the author)

Figure 8.8:
An example of Rockmart's harder slate, still in good condition despite its age.
(Photo by the author)

Blance was born on May 8, 1799, in Savannah, Georgia, of French parents, and moved to the Rockmart area in 1847. One story describes him stumbling over an outcropping of slate as he was tying up his family cow, thereby making his memorable discovery. Another version describes Mr. Blance's discovery as a byproduct of a road-grading project he was doing at the time (maybe the cow was pulling the grader). In any case, he subsequently established the Blanceville Slate Mines and invited the Welsh to step in and lend a hand. No doubt the Welsh were lured by stories of *"the largest, most inexhaustible deposit of slate south of Pennsylvania"* reported discovered near Van Wert (an extension of Rockmart). One notable Welshman was Colonel Seaborn Jones, credited as one of the originators of the Rockmart slate industry. Colonel Jones established extensive land holdings and became a major benefactor of the city of Rockmart.

The average value of farmland in Polk County in 1850 when slate was being discovered, was about $10 an acre, and most people earned their living growing cotton, corn, peaches and other fruits, or operating dairy farms. Some worked in cotton mills. That all changed as the slate industry expanded and became a major employer of men in the area.

Although the Civil War interrupted the development of Rockmart's slate industry as the men took up arms and engaged in battle, in 1863 some mining was done to provide roofing for military purposes. The quarries nevertheless remained at a standstill for fifteen years after the war, reopening around 1880. They then produced their greatest yield during the following decade, with 5,000 squares of roofing slate (enough to cover 420 houses) valued at $22,500, being produced in 1894. According to the U. S. Geological Survey, Georgia produced 38,097 squares of roofing slate from 1879 until the beginning of World War I, valued at $165,918, although these figures are thought to be incomplete. It is reported that the total roof slate production of the Rockmart district was over 50,000 squares, enough to cover about 4,200 average size homes.

Roofing slate production dropped off after 1900 and was replaced by the quarrying of slate to make light weight aggregate. The aggregate, known as "Galite" (a contraction of Georgia lite), was made by heating crushed slate to 2000 degrees F., which causes it to expand. The resulting "expanded shale aggregate" was added to concrete to reduce the concrete's weight, but not its strength, and is incorporated into such structures as the Golden Gate Bridge and the White House. The Rockmart Slate Corporation quarry in Rockmart, Georgia, was planning to renew the production of ground slate in 1997 to be used as a horticultural mulch material called "Slatescape."

FAIRMOUNT

The slate near Fairmount, thought to be Cambrian in age (600 million years old), is green, similar to the sea green slate of Vermont, but with more color variety, showing shades of blue, buff, brown and red. Fairmount slate rock at quarry sites appeared to the author to be quite hard, although not a single Fairmount slate *roof* could be found in the area by the author and a colleague, who spent a couple of days in 1997 searching the Fairmount area for slate roofs, in vain. It is reported in historical accounts that these slates did not "possess the hardness of the best Rockmart slates," and due to the lack of existing roofs, the author had no way to visibly and with certainty determine the longevity of Fairmount slate roofs.

There weren't many local folks who remembered the operating slate quarries, and those that did only remembered the slate being ground into aggregate. No one was old enough at the time to remember slate ever being quarried for roofing, although one older farmer did remember a hotel on Slate Mine Road that had Fairmount slate on the roof, but which had burned to the ground. It appeared that, although Fairmount slate was once quarried for roofing, it was only done in small quantities, and the roofs were eventually lost to neglect or ruin. There are some roofs in Atlanta, and presumably in other areas of Georgia, with green slate roofs which may have originated in Fairmount, but the origin of the roofs is pure speculation due to the similarity in appearance to green and sea green roofs of Vermont.

The green slate belt was first prospected by Mr. G. W. Davis, a slate worker from Pennsylvania employed in the Rockmart district. The exact date is uncertain. The belt extends through Gordon, Bartow and Murray Counties. Huge water-filled pits remain hidden among the forested hills about four miles south of Fairmount, bearing evidence of the old slate industry. From these pits emanate large tunnels for a total distance underground of 12 miles, according to quarry owner Ray Sullivan (see Figure 8.10).

Figure 8.9
▲ **Slate Mine Road near Fairmount, GA, is reportedly the site of the old shingle mines, but only traces remain there. The huge quarry pictured on the next page is several miles away, on the other side of town.**
(Photo by the author)

Figure 8.10:
▲ Abandoned slate quarry pit near Fairmount, GA. Note person standing near top, just right of center.
▼ These huge tunnels extend 12 miles underground, now only accessible by boat. (Photos by the author)

Figure 8.11:

▲ **Typical Rockmart, Georgia, slate roof, somewhat refurbished.**

(Photo by the author)

References - Chapter 8

VIRGINIA

• Bridgen, John; *"Turning Stone into Bread,"* article of unknown origin and date, pp. 10-14.
• Buckingham Slate Corporation BuyLine 0459; January, 1990, and other published information, Buckingham-Virginia Slate Corporation, One Main Street, Arvonia, VA 23004-0008 (1994).
• Chambers, S. Allen, Jr., (1989); *"Of the Best Quality - Buckingham Slate,"* article in "Virginia Cavalcade," Spring 1989, pp. 158-171.
• Lindsay, Jean, (1974); A History of North Wales; David and Charles, Newton Abbot, London; p. 254.
• Redden, J. A., (1961), *"Slate in Virginia,"* article in "Mineral Industries Journal," Vol. VIII, No. 3, September, 1961, published by the Virginia Polytechnic Institute, Mineral Industries Departments, School of Engineering and Architecture, pp. 1-5.
• Terrell, Patricia, (1962); *"Virginia Slate Industry: An Old Reliable Shows New Vigor,"* article in "The Commonwealth," December 1962, p. 110.
• Tucker, Beverley R. Jr., (date unknown); *"Slate, Past Present and Future,"* Buckingham-Virginia Slate Corporation, Richmond, Va.

GEORGIA

• *Galite Lightweight Structural Concrete, A Basic Manual*; Georgia Lightweight Aggregate Company, Atlanta, GA.
• Geological Survey of Georgia, Bulletin 27, 1912.
• Knight, Lucian Lamar, (1917); Georgia and Georgians, Vol. II; Lewis Publishing Co., Chicago and New York.
• Memoirs of Georgia, Volume 1, (1895), The Southern Historical Association, Atlanta, GA, p. 206.
• Mintz, Leonora, *"Slate Discovery Brought Mines to Van Wert"* (newspaper article found in Rockmart, Georgia, Public Library; no date or source accompanied article). Other photocopied articles from the same library, kept in a manila file folder in a back room, also devoid of dates, page numbers or sources, were also used in the section on Georgia slate in this chapter.
• Sargent, Gordon D., and Jackson, Olin; "The Town that Stone Built"; *North Georgia Journal*; Summer, 1996; pp. 28-35.
• *Some Historical Facts About Rockmart's First Industry - Slate*; The Rockmart Journal, Thursday, July 20, 1967; p. 7-A.

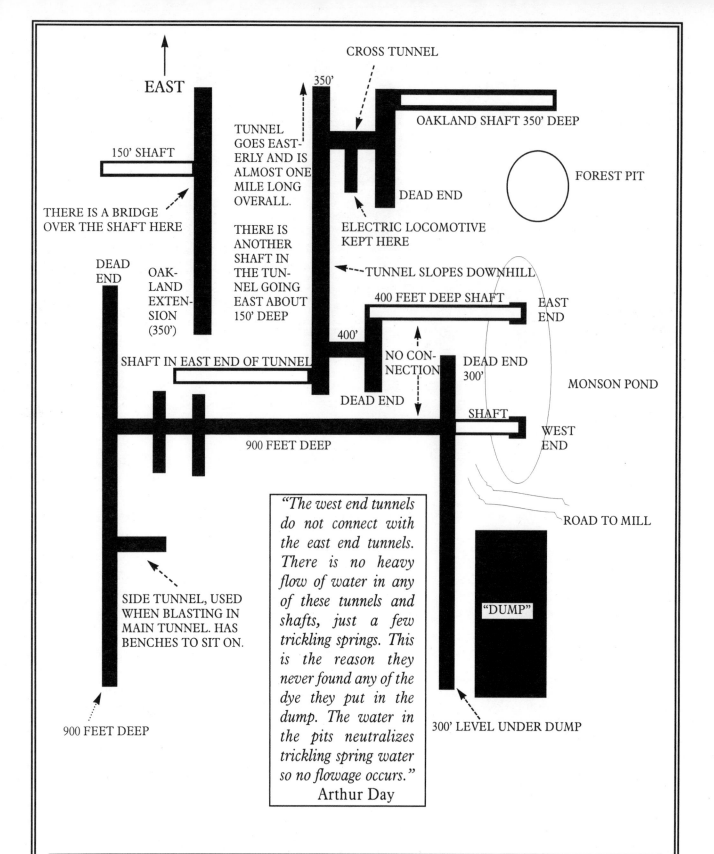

EAST

CROSS TUNNEL

350'

OAKLAND SHAFT 350' DEEP

150' SHAFT

FOREST PIT

THERE IS A BRIDGE
OVER THE SHAFT HERE

TUNNEL
GOES EAST-
ERLY AND IS
ALMOST ONE
MILE LONG
OVERALL.

THERE IS
ANOTHER
SHAFT IN
THE TUN-
NEL GOING
EAST ABOUT
150' DEEP

DEAD END

ELECTRIC LOCOMOTIVE
KEPT HERE

DEAD
END

OAK-
LAND
EXTEN-
SION
(350')

TUNNEL SLOPES DOWNHILL

400 FEET DEEP SHAFT

EAST
END

400'

SHAFT IN EAST END OF TUNNEL

NO CON-
NECTION

DEAD END
300'

MONSON POND

DEAD END

SHAFT

WEST
END

900 FEET DEEP

ROAD TO MILL

SIDE TUNNEL, USED
WHEN BLASTING IN
MAIN TUNNEL. HAS
BENCHES TO SIT ON.

"The west end tunnels
do not connect with
the east end tunnels.
There is no heavy
flow of water in any
of these tunnels and
shafts, just a few
trickling springs. This
is the reason they
never found any of the
dye they put in the
dump. The water in
the pits neutralizes
trickling spring water
so no flowage occurs."
Arthur Day

"DUMP"

900 FEET DEEP

300' LEVEL UNDER DUMP

A MONSON SLATE QUARRY

DIAGRAM OF MINE SHOWING SHAFTS AND DETAILS

From a hand drawn map by Arthur Day, slate quarry worker for Monson, Maine Slate Co., drawn at about age 80, from memory. (Special thanks to Berwin Storer)

Chapter Nine
Maine (and Canada)

The slate from Monson, Maine, is black, smooth and will last for centuries on a roof. The people of that area will say what is commonly heard in all slate regions in the USA, that their slate is "the best slate in the world." Unfortunately, at the time of this writing, Monson slate is no longer being quarried for roofing.(*see note on page 142). Quarries stopped producing roofing slate in Monson for the same reason they stopped elsewhere throughout the world: increased cost, market fluctuations and decreased demand.

Older Monson slate roofs are, nevertheless, excellent roofs very much worthy of restoration efforts. People who put their money or efforts into restoring their Monson roofs will benefit greatly in the long run, both financially (by keeping a permanent roof over their heads) and aesthetically (by keeping a traditional, natural, stone roof over their heads).

The restoration of Monson roofs requires ample supplies of replacement slates, and since the slates are no longer being manufacturered, old roofs that are being removed (as on buildings to be demolished) should be carefully salvaged and stored until needed. This careful recycling of Monson slate will ensure that many of the old roofs will be kept intact, as one of the most necessary ingredients in slate roof restoration is replacement slates that match the slate on the roof. Recycling of roof slate is discussed in greater detail in Chapter 12.

Monson is a quiet, quaint little village, founded in 1822, just two years after Maine gained statehood. Nestled on the shore of Lake Hebron, near Monson Pond, it is just a few miles south of the beautiful Moosehead Lake in north central Maine.

Life changed for the residents of Monson in the summer of 1870, when William Griffith Jones, a Welshman, discovered an outcropping of slate while riding his horse. Welshmen at that time were considered the foremost authorities on excavating and cutting granite, marble and slate, and apparently Mr. Jones understood that slate had commercial value, so he bought the slate-bearing land from a hotel keeper and started the development of his Eureka slate quarry within days.

Mr. Jones discovered the tip of an iceberg when he first saw that slate outcropping, for it was a tiny section of one of Monson's two immense slate belts. The village belt was 200 feet in width, and became one of two belts known as the Monson-Maine Co. vein and the Portland-Monson Co. vein, both of which ran east to west.

Soon after Jones' discovery, slate quarries sprung up all over the area. The first quarry after Jones' was called the Hebron Pond Quarry, started by a Mr. Chapin. It suffered an unfortunate accident in 1881, when two men were injured and a third man killed. In 1872 a Mr. Norris opened a quarry and installed steam power, a new-fangled addition to the Monson area quarries

FIGURE 9.1

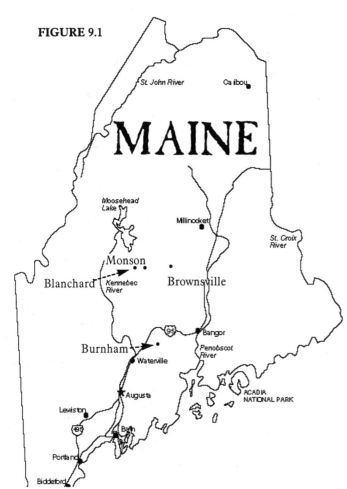

at that time, but he didn't have much luck with it - his buildings burned to the ground five years later.

In the meantime, many people of English, Irish and Scottish descent immigrated into the Monson area from New Brunswick and Nova Scotia to develop and work the quarries. William Thomas, Commissioner of Immigration, brought Swedes to the area in 1870, and later Finnish people arrived to help work the slate mines.

In 1873 Fred Jackson founded the Cove Quarry, which utilized some of Mr. Norris' machinery, thereby sparing their loss in the later fire. The same year, Dexter and Portland Companies founded the Dirigo Quarry, followed by the Forest Quarry (opened by people named Forest) in 1874. Then the Underground Quarry, operated by John Folsom sprung up in 1875; then John Tripp's Oakland Quarry in 1877; the Monson-Maine Quarry in 1880 (by Allen Williams); the Kineo Quarry and the Burmah Quarry in 1882; the West Monson Quarry (W. M. Jones) in 1895 (the buildings burned January 16, 1897); the Matthews Quarry in 1902; the Portland-Monson Quarry in 1906; the Farm Quarry and the Wilkins Quarry in 1910; and the Eighteen Quarry in 1919 (abandoned in 1922).

Some of these smaller quarries were eventually bought up by the larger ones, and by 1882, 12 years after Mr. Jones discovered the first slate outcropping, $150,000 had been invested in the operation of five quarries in the Monson area. Six hundred and twenty carloads of roofing slate equivalent to 2,500,000 square feet of roofing were being manufactured annually, requiring 200 men and 35 horses, as well as twelve teamsters to haul the slate to the railroad. Two thousand five hundred cords of wood were used each year, and about $6,000 was being generated annually in local wages by the quarries. By 1885, a newspaper was being published out of Bangor, Maine, called "The Monson Slate."

By 1922, a scant forty years later, there were only three quarries in operation - the Portland-Monson, the Monson-Maine, and the General Slate Co. The Monson-Maine Quarry outlived the General Slate Co., but was eventually abandoned in 1943. The Portland-Monson Slate Company was run by the Coleman family until April 1, 1965, eventually to change hands to the Tatko family, who renamed the company the Sheldon Slate Products Co. in the early 1990's. Sheldon Slate Products of Monson, Maine, the only slate quarry in operation in that area today, produces countertops, floor tiles, flagstones etc., but not roofing slate.

At the beginning of this chapter is a map of the Monson-Maine Co. mine shafts and tunnels as recorded by Arthur Day, a quarryman who drew the map for posterity in the 1980's at about age 80. He indicates that some of the tunnels descended to 900 feet, and that an electric locomotive eventually came into use in the tunnels, presumably to move the slate along tracks laid on the tunnel floor. He also shows where a side tunnel was carved from the main tunnel to

Figure 9.2: ▲ This house in Monson, ME, has a Monson slate roof. The slate pieces in the foreground are for sale. ▼ An abandoned Monson slate quarry pit, hundreds of feet deep, has filled with water over the years - a typical sight in slate mining areas of the United States.

(Photos by the author)

allow men a safe place to duck, complete with benches to sit on, when blasting took place in the main tunnel. Other quarrymen of the area tell of mine tunnels 1100' deep, and of a shaft under Monson Pond, two of which are evident in Mr. Day's diagram.

In addition to the quarries in the Monson area, the neighboring town of Brownsville, twenty miles to the east, had quarries that produced much of the famous roof slate known as Monson slate. Also, a small town called Blanchard a few miles to the west of Monson quarried slate, as did Burnham, about 45 miles south of Monson (see Figure 9.1).

The quarry business was not without its dangers, as we have seen in previous chapters, and Monson was not immune to them. Numerous accidents took place over the years in addition to the loss of buildings to fire. Not only were workers killed or injured in the mines, but people drowned in abandoned quarry pits that filled with water, and others died after falling over the cliff-like sides of the open slate pits.

Perhaps the most well-known use of Monson slate was for the Kennedy Memorial stone marking President John F. Kennedy's grave at Arlington National Cemetery. Saint Patrick's Cathedral in New York City also has a Monson slate roof.

*A roof slate very similar to Monson slate is now (1997) being produced at the Glendyne Quarry in St-Marc du Lac Long, Quebec, Canada, just north of the Maine border. It's available through Newfoundland Slate, Inc. (see page 218).

(Much of the historical information about Monson's quarries comes from *"The History of Monson, Maine - 1822 - 1972,"* published by the town of Monson, Maine).

Figure 9.3
▲ **MONSON SLATE ROOF**
Note the smooth texture and sheen on the hard, long-lasting black slate of Monson, Maine.

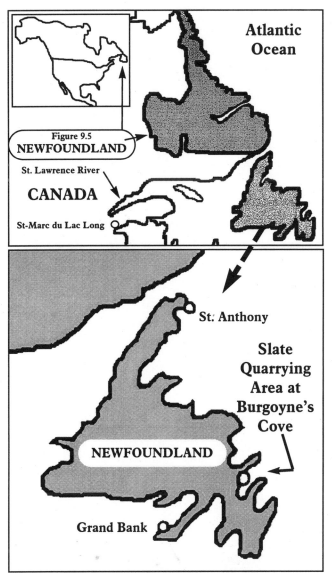

Figure 9.5
NEWFOUNDLAND

Atlantic Ocean

St. Lawrence River

CANADA

St-Marc du Lac Long

St. Anthony

Slate Quarrying Area at Burgoyne's Cove

NEWFOUNDLAND

Grand Bank

CANADA

"Newfoundland possesses one of the great roofing slate deposits in the world."
Professor G. B. Walcott, Director of the Smithsonian Institute (late 1800's)

One of the first known slate quarries to be opened in Canada was near Rimouski on the south bank of the St. Lawrence River. Opened in 1728, the slate was found to be of a poor quality and the quarry was abandoned in 1733.

A once productive slate district is located about 60 miles north of Newport, Vermont (USA), near Richmond and Danville, Quebec. Here, slate was first mined in 1854 at the Steele Quarry (later called the Bedard Quarry), about 3 kilometers southeast of Richmond. The first profitable quarry opened in 1860 less than 2 kilometers northeast of the village of Kingsbury, and was known as the Melbourne Quarry (also called Walton Slate). A square of slate at that time sold for $3.50 Canadian, while the quarry wage 30 years later still only hovered around eighteen to twenty-three cents an hour.

In the last few decades of the 1800's, Melbourne District quarries provided most of the slate used for roofing in Canada. In 1878, the Melbourne Quarry shut down, and the New Rockland Quarry, which began in 1868, then became the only important slate quarry in the region, monopolizing 80% of the Canadian roofing slate market. The New Rockland Quarry employed 150 - 300 workers in the late 1800's, many of whom were Welsh and Cornish, and who were localized in the village of New Rockland. The quarry continued until 1924, after which the inhabitants of the village moved away and now no buildings remain there. By the time the New Rockland Quarry was in its waning years, it didn't produce more than a third of the slate consumed in Canada, while much was instead imported from the United States.

Today, one may visit Melbourne's "Slate Interpretation Center," a slate-roofed museum at 5 Belmont Street, Melbourne, Quebec, just across the river from Richmond via an old steel bridge (10 miles from Kingsbury, or about 60 miles north of the Vermont border). The center was incorporated in 1992, and is dedicated to the preservation of slate roofs in the Kingsbury-Melbourne-Richmond area of Quebec.

Slate was discovered in the mid-1800's on the eastern seacoast of Canada in Newfoundland, the slate being of Cambrian age, meaning it had been highly metamorphasized into a very hard and durable stone, and is therefore categorized as S1 (hard) slate. This type of slate can be expected to last hundreds of years on a roof, with proper maintenance. It is identical to the slate of Wales, particularly from the Penrhyn, Dinorwic, and Dorothea Quarry areas, yielding shades of purple and green.

The first quarry in the Newfoundland region was opened by the Welsh in the mid 1850's and most of the slate quarried at that time was used locally or shipped overseas. The quarry shut down in the early 1900's, as did many others throughout North America, due to economic hardships and market fluctuations. However, the Newfoundland quarry *re-opened* in the mid-1980's, and in 1991 the ownership changed and the operation expanded into a new, large mill building with new equipment. Today, roof slate is produced there under the trademarked brand name "Trinity Slate" using fully automated equipment including diamond wire machines in the quarry, laser guided saw systems for cutting blocks, pneumatic hammers for splitting the individual roof slates, and pneumatic trimmers for trimming the slates to their final size.

Newfoundland, Canada, now boasts the largest production of natural roofing slate in North America with distribution in the UK, Europe, Japan, Australia, New Zealand, and North America. Newfoundland Slate, Inc. with its quarry site at Burgoyne's Cove, off Trinity Bay, lays claim to being "the only slate quarrier in North America with the prestigious ISO 9002 Certification," an international standard for quality assurance.

Roof slate is also being quarried at the Glendyne Quarry at St-Marc du Lac Long in Quebec, which is about a hundred miles north of Monson, Maine. This quarry began operation in 1995 and yields a slate comparable in quality and appearance to the famous Monson Slate of Maine. Glendyne slate can be purchased through Newfoundland Slate, Inc. in Scarborough, Canada (the reader may refer to page 218 for information about how to contact this, and other, sources of roof slate).

Most of the slate roofed buildings in Canada are located in Toronto, London, Brantford, and Montreal. More than two-thirds of these are on private residences, especially on upper-middle class homes. Slate roofs are also found in the various slate quarrying regions, especially near the old quarry sites.

[Information on the history of Canadian roof slate production came from Trinity Slate Co., Inc; the Centre d'interpretation de l'ardoise de Kingsbury (Slate Interpretation Center); and the Slate Valley Museum, Granville, NY]

Figure 9.6: ▼ Trinity Slate in Newfoundland splits its roof slates with pneumatic splitters.
(Photo courtesy of Newfoundland Slate Co. Inc.)

CHEMICAL COMPOSITION
(Tests may vary)
Newfoundland Slate (%?)
Silica 58.85
Alumina 19.65
Iron 8.42
Potassium 3.75
Magnesia 2.30
Sodium 1.72
L.O.I. 3.30
Trace 1.80

ASTM Average Properties

	Purple	Green
Modulus of Rupture		
Across Grain	9,196 -	10,906
Along Grain	6,417 -	6,417
Absorption	0.15 -	0.10
Acid Resistance	0.0005 -	0.0002

(Source: Trinity Slate Doc. QD 10/3-5a Issue 1)

▲ This magnificent Peach Bottom slate roof, installed in 1893 on the Buhl Mansion in Sharon, PA, is shown here undergoing restoration. The mansion is being converted into a guest house, art gallery and spa.

▼ Rounded valleys and dormers provide an elegant look more common in Europe, seen here on the Ketler home in Grove City, PA. The slate is Vermont green, sea green, and purple, in a graduated pattern. Photos by the author.

▶ A worker prepares a Vermont sea green roof for salvage by first removing ridge iron and lightning rods. Although this Pennsylvania barn was demolished, all of the slates were salvaged and recycled onto new buildings.

(Opposite) Welsh slate walls and roof of the Llanberis area near the Dinorwic Quarry. Welsh slate is very similar to Vermont and Canadian slate in color and durability.

▼ A Vermont green slate roof with New York red slate design, on a house in Vermont.

(Photos by the author)

The Slate Roof Bible

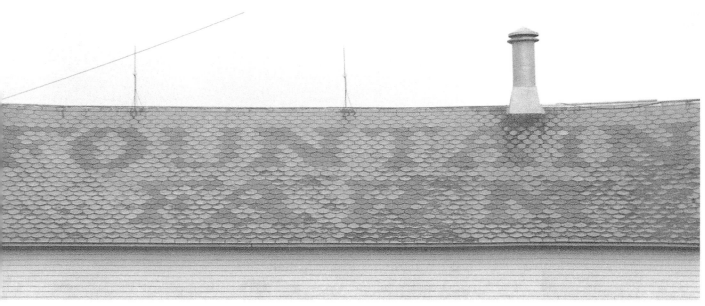

▲ Classic roof design on a single sea green Vermont slate barn roof near Townville, PA. New York red slate inscription reads: P. E. Wood 1900 Fountain Farm.

▼ Two scenes of Welsh roof tops, nearly all are slate. Left is Barmouth, right Dolgellau. (Photos by the author)

▲ Very hard, durable, red slate from Granville, NY, area quarries - shown on a church in Foxburg, PA.
▼ Vermont purple slate with NY red slate and VT green slate design on church in Poultney, Vermont.

(Top photo by Barry Smith, bottom photo by the author.)

▲ Century old, recycled "sea green" slates installed on a new building at Slippery Rock University's Harmony Homestead, Slippery Rock, PA (see also page 213). The lettering is done in Vermont purple slates.

▼ Recycled VT sea green and VT purple slates on a new garage near Plain Grove, PA. Recycling roof slate is discussed on page 211.   (Photos by the author)

▲ This chicken coop roof is made of recycled Vermont sea green slate with a Pennsylvania Bangor black slate design. ▼ Although most sea green roofs appear light gray with reddish slates scattered throughout, as above, some sea green slate weathers to a very dark shade, as shown on this Knights of Columbus building in Sharon, PA. (Photos by the author.)

Foreword - Part II

What's It Like to Work on a Slate Roof?

A friend once asked me what it was like to do slate roof restoration work. He worked in an office so I decided to describe my work in terms he would understand:

Imagine you're in your office one fine summer day, sitting at your desk, comfortably typing away at your computer, listening to the hum of the air conditioner while your wonderful co-worker graciously brings you a freshly brewed cup of coffee.

Then the air conditioner dies. You continue typing until you realize that not only has the air conditioner stopped, but someone has turned the furnace *on,* and it's summer! As the temperature rises and rises, passing 100^0F and climbing, the sweat beads up on your brow and you become aware that someone has also turned *on* a *humidifier.* The heat is so intense and muggy that soon you start ripping off clothing, flinging your suit coat and tie to the floor, along with your shirt. Sweat is now pouring into your eyes, blinding you, so you grab a rag and wrap it around your forehead. Your co-worker informs you that this is now your new work environment, and you must do your day to day tasks under these sweltering conditions.

Stunned by this revelation, you pull your desk drawer open, clumsily groping for some paper clips, and a cloud of black, sooty dust flies out in your face covering your sweaty body. You now look like a coal miner as you take a tissue and wipe the dirt granules out of your eyes so you can continue to work. You pull your drawer open further and a swarm of yellow jackets flies out. They're obviously mad as they go for your face. Some lodge in your hair as you panic and fall backward in your chair, scrambling to get away.

You run for the closet to get a can of hornet spray, and as you poke around looking for the spray, swatting at the hornets in your hair, a bat flies out of nowhere, startling you. You duck as the bat skims past your head chirping like crazy, then you watch it assume a circling pattern in the office as it flies around looking for a place to hide. Fortunately, you're not afraid of bats or hornets, but when you grab the can of hornet spray you yell and drop it on the floor. The temperature has risen to 135^0F and the can is too hot to touch. You pick up a pen and drop it too, for the same reason. So you put on some gloves, despite the heat, grab the spray, sneak back to your desk and spray the hell out of the yellow jackets, while muttering some vengeful obscenities. Then you fling open a window to get some air and, luckily, the bat flies out. Unfortunately, when you flung open the window another cloud of soot flies at you, but this time you're ready and you close your eyes before the dirt hits them.

Now that the bees have been killed (except for the ones who were away when you sprayed and are returning to what's left of their nest), you sit back down at your desk, being careful not to touch anything metallic without insulating yourself from the hot surfaces. Only now you find that your typewriter is gone, and in its place are steel and iron hand tools. Your co-worker informs you that your job henceforth will be to beat on these tools as hard as you can with a hammer. "And don't cut yourself on the sharp edges!"

"Sharp edges of what!?" you cry.

"Of the slate and sheet metal, of course!" is the impatient reply. "They can be as sharp as razor blades, dummy!"

You immediately nick a finger the first time you impatiently beat on a slate ripper, so you stop and wrap the bleeding appendage with a bandage, wondering if maybe you should have washed your black hand first. When you resume your labor you realize that your desk is covered with a thick layer of bat droppings, and you've been grinding the guano into the skin of your arms. Just when you've stopped to ponder this new discovery an amazing and unexpected thing happens, the floor of the room heaves up, tilting at 45 degrees! Everything in the room goes flying - downhill! Your desk remains attached, however, and you cling to it for dear life.

Your co-worker yells up to you from the bottom of the heap that she has a hook ladder you can hook to the top of the room so you have something to stand on. She slides the ladder up the slanting floor past your desk and somehow it hooks at the high end of the room. You carefully stand on the ladder, crouched low and leaning to compensate for the 45 degree angle of your new work surface. "Back to work!" she yells up, "but here, take these." She throws up a pair of rubber soled canvas shoes. You slowly take a seated position on the hook ladder, and remove your wing tips. The soft, lightweight shoes give you better traction.

Back to work!? OK. But you find that the "floor" is so hot you can't touch it and you can't even rest your foot on it to stabilize yourself because it burns the soles of your feet right through the shoes. You must keep your entire body perched on the aluminum hook ladder, which holds you an inch or two off the floor surface, now radiating heat like a heat lamp. You take off your sweat band and wipe your sweaty forehead, clinging with one gloved hand to the ladder. Then you replace the sweatband and grab the iron implements. The last thing you're worried about right now is bats or bees, although you don't want to be startled under these conditions and risk falling. So you move slowly and deliberately as you once again begin beating as hard as you can on the iron tools at your desk.

But you can't relax yet. Suddenly, the entire work environment lurches forty feet in the air and all the walls and ceilings drop away leaving you out in the beating sun on a steep slope with your hind-end flapping in the wind. The heat, dirt, bats, bees and humidity remain with you, however, and although there is an occasional breeze, it doesn't seem to do anything but blow the hot air around. With both hands clinging to the ladder, you look over your shoulder and see your co-worker standing on the ground. "No more pushing paper for you," she yells up. "You need to come down here and carry up these slabs of stone!"

"What! How many?"

"Oh, a ton or so," she says, without exaggerating.

You are now asking yourself if your miserable working conditions can possibly be any worse. You can't imagine what else can go wrong when suddenly you smell something, and it doesn't smell good. "Oh god, what's that?" you ask yourself outloud. "What's what?" replies your co-worker in a gruff voice. You're startled to find that she is now clinging to the same hook ladder you're clinging to, and she's had a sex change operation. She now bears a striking resemblance to a caveman. Worse, she has developed a notable proclivity for the use of expletives in her language. The smell? As luck would have it, your co-worker is now listed in the Guiness Book of World Records as the world's greatest source of natural gas!

Just at that moment, a thunderstorm blows in out of nowhere and lightning begins to flash. You realize you've become a human lightning rod and you scramble down the hook ladder to get off the roof. Rain begins to pelt down and the wet roof surface turns as slick as ice. Your co-worker has enough sense to get down ahead of you, so you join him on the ground where you seek shelter together. The pouring rain has cooled things off, washed some of the dirt off your face, and now you've got your feet back on the ground. Your prayers of gratitude are suddenly interrupted with a shocking thought - you forgot to close the gaping hole you left in the roof!

[Unfortunately, the above scenario is not (much of) an exaggeration.]

Chapter Ten
Safety First

ost old slate roofs are in need of repair, and ninety percent of what's involved in repairing them is just a willingness to climb. If a slate roof could be laid down on the ground, almost anyone would be willing to get on it and repair it. But it's not down here - it's up there. This is where the element of risk comes into play, and safety precautions must be taken seriously.

You can actually walk out the front door of your house, trip on your shoe string, fall, hit your head on a rock and die. If you fall from a height of six feet, same story. If you fall from a roof top . . . well, let's not imagine the consequences, but let's be aware of the possibility and take measures to prevent any sort of accident. The author, at the time of this writing, has worked on slate roofs for nearly 30 years and will try to pass on some insight and experience related to safety when working on the high, steep, slippery surfaces characteristic of slate roofs.

But first the obligatory disclaimer. I'm passing on this information for educational purposes, but I can't be held responsible for the actions of others. The author cannot control the attitude of the reader, nor sharpen his or her senses. You are responsible for that, and if you trip on your shoestring on the way to the ladder, don't blame me. When working at heights, you must proceed with utmost care and caution, and *you* must take full personal responsibility for *yourself*. Read all instructions carefully, and follow the recommendations of the manufacturer of any roofing tools and/or equipment. Be aware of safety codes and regulations in your area, and do not proceed with any roofing work unless and until you, and you alone, will take full responsibility for your actions.

Perhaps you heard the true story of the man who wanted to work on his roof but feared for his safety. He tied a stout rope to the rear bumper of his car, flung the rope up over the peak of his house, then climbed up a ladder on the other side of the house and tied the rope around his waist. While he was working happily away, his wife had a sudden and unexpected urge to go grocery shopping. Unfortunately, the man had neglected to inform his wife of his ingenious safety precautions. She hopped in the car and took off without a second thought. This story was reported in the newspaper, and although the man was yanked off the roof and survived, his final condition, although unfortunate, was unclear.

Even though this man had safety in mind, he went about it in the wrong way. This illustrates a case where a person violates all the rules of safety - he had the wrong attitude (too careless), didn't understand the potential hazards (in this case, not informing his wife), and he didn't use the proper equipment (ropes wrapped around one's waist are a mistake).

[The following page illustrates types and parts of roofs with which you should familiarize yourself in order to best understand the rest of this chapter, as well as the rest of this book.].

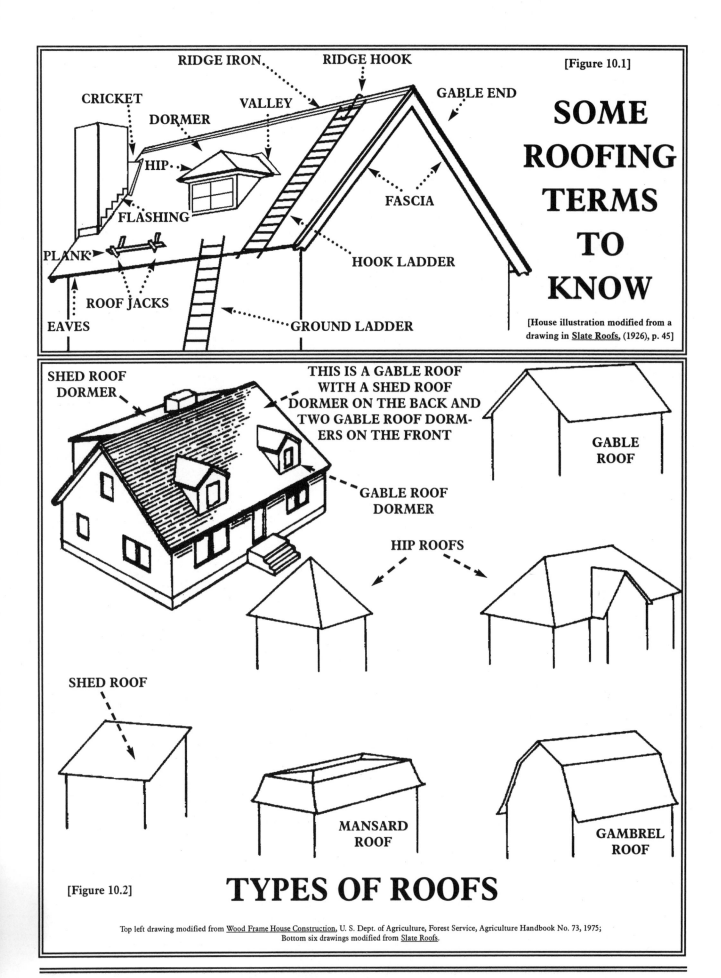

[Figure 10.1]

SOME ROOFING TERMS TO KNOW

CRICKET

RIDGE IRON

RIDGE HOOK

DORMER

VALLEY

GABLE END

HIP

FLASHING

FASCIA

PLANK

HOOK LADDER

ROOF JACKS

EAVES

GROUND LADDER

[House illustration modified from a drawing in Slate Roofs, (1926), p. 45]

SHED ROOF DORMER

THIS IS A GABLE ROOF WITH A SHED ROOF DORMER ON THE BACK AND TWO GABLE ROOF DORMERS ON THE FRONT

GABLE ROOF

GABLE ROOF DORMER

HIP ROOFS

SHED ROOF

MANSARD ROOF

GAMBREL ROOF

[Figure 10.2]

TYPES OF ROOFS

Top left drawing modified from Wood Frame House Construction, U. S. Dept. of Agriculture, Forest Service, Agriculture Handbook No. 73, 1975; Bottom six drawings modified from Slate Roofs.

SAFETY
FIRST

1. KEEP A SAFETY ATTITUDE
2. KNOW THE HAZARDS
3. USE PROPER EQUIPMENT

1) PROPER ATTITUDE

First, and perhaps most important, safety is a mental attitude. You must be careful and cautious. You must realize that *you* are responsible for your own safety, and if you fall or hurt yourself *you,* and perhaps your family as well, must suffer. If you don't take care of yourself, no one is going to do it for you. You probably understand this when driving a car or when chopping vegetables at the kitchen counter; you *must* understand this when climbing on a roof. The man in the incident above could have at least opened the hood of the car to indicate that it shouldn't be moved (or he could have put a note on the steering wheel). You don't want to be telling yourself what you *should* have done - instead, think ahead!

Never rely on someone else's judgement about whether a situation is safe or not (unless you're an apprentice working *in person* alongside a *master* roofer). Use your own judgement. If you don't feel safe in any given situation, don't proceed. Stop and do whatever is necessary to improve the safety of the situation to *your* satisfaction, then continue.

Never be in a hurry when working on a slate roof. It only takes one slip and you're history. This issue about attitude cannot be overemphasized. Keep your eyes and ears open all the time when working at heights.

Always make certain the correct equipment is being properly used and *always* be aware of the potential hazards. Find them before they find you, and prevent accidents before they happen.

2) BE AWARE OF THE ROOFTOP HAZARDS

There are several hazards that exist on old slate roofs that you must memorize and look for *every time* you get on *any* roof. They can be summed up as: BEES, BATS, CHIMNEYS, WIND, LIGHTNING, ELECTRICAL WIRES, and CO-WORKERS.

BEES

Bees are one of the worst hazards on old slate roofs. One recent summer in a local town a man was found at the bottom of a ladder riddled with bee stings, and dead. He had probably set his ladder up in front of a nest of yellow jackets, which may not have been apparent to him, and

Yellow Jacket

Honey Bee

Carpenter Bee

Mud Wasp

White-faced Hornet

BEES ARE ONE OF THE MOST SERIOUS HAZARDS WHEN WORKING ON SLATE ROOFS

when they attacked him he panicked, fell off and thereby made his last mistake. You must *always* look for nests of bees during the summer months on any house, rural or urban, before even putting up a ladder. There are generally seven types of common bees: yellow jackets, honey bees, paper wasps, bald-faced (white-faced) hornets, bumble bees, carpenter bees, and mud daubers.

The first and worst are **yellow jackets**. These are small, mean and nasty wasps that live in highly populated paper nests sometimes shaped like footballs, but the nests are usually *not* visible - they're in the wall or roof of the house. Sometimes the nests are in the ground and yellow jackets are sometimes called ground bees (there are other types of ground bees too). Sometimes the nests are located (exposed) under the eaves of the building anywhere from down on the porch to the very top of a three story house. Yellow jackets look sort of like honey bees and many people can't tell them apart, but honey bees are fuzzy and yellow jackets aren't. Also, yellow jackets die off in the winter *except for the queen*, who hibernates and starts a new nest in the spring. Therefore, yellow jackets don't nest in the same spot from one year to the next, but they *will* nest in the same house. Honey bees, on the other hand, usually stay put in the same nest year after year.

One reason yellow jackets are such a hazard is because they're just plain *mean*. They'll go out of their way to sting you. I've had them fly from under the eaves clear over the peak of the house to the other side of the roof just to try (successfully) to sting me. When you're hanging on a ladder and perched precariously at some considerable height, you don't want a bunch of yellow jackets flying in your face. It's a natural reaction to recoil, swat like hell and run when attacked by bees (or is it recoil, swat, and run like hell?). Either way, you can't do that when on a ladder or a roof, so you want to make sure you don't put yourself in that situation. If by chance a bee does fly at you it is critical that you hang on to whatever it is you're holding on to, and don't panic! Leave the area as quickly as possible without being reckless. Don't let a puny bee force you to meet your maker.

I might add that some people are allergic to bee stings, and yellow jacket stings are the most venomous, according to my observations. A friend of mine was stung in the forehead by one and her head swelled up like a pumpkin. She looked like something out of a horror movie for about two days. My neighbor was stung by one and his whole body swelled up and he had trouble breathing for hours. I have worked on slate roofs since 1968 and have never been stung by any bee except yellow jackets, and only once or twice. I am not allergic to them and was once stung by 17 of them as a teenager (not on a roof) with no adverse reaction. But now I spot them before they spot me and I avoid them if I can. If I can't avoid them I spray them with hornet spray *if* I can get close enough to spray the nest itself. It does no good to just spray the bees because there are thousands of them, but if you can knock out the nest the remaining live bees will likely leave you alone (they become demoralized and disoriented after the queen has died).

Being a beekeeper myself (honeybees) I can safely say that the best defense against harmful bees is awareness of their habits, patterns, and weaknesses, avoidance of them, then eradication if all else fails. I don't recommend hunting down and killing every yellow jacket nest you can find, because then you're just asking for trouble, but if you do need to kill them here are some tips:

Most of the bees will be out working on a sunny day. That's how you can locate the nest - look for the insects flying toward and away from the house. When you see insects flying back and forth like that, you can trace their flight path back to the house and either find the nest, or you may see a small hole under a window sill, between some clapboards, in a chimney, or even on the roof itself where a slate is missing, and you'll see the bees flying in and out of the hole. Once you've located the nest (almost *every* old farmhouse and many city houses have them), stay away from it. *Do not* put a ladder up in front of a bee flyway. Remember that bees, like some people, are friendliest on sunny, dry days, and meanest on cloudy and muggy days. This is a fact. So on a sunny, dry day you may be able to work closer to a bee nest than on a bad weather day, but don't count on it.

Because the bees are working during the day, the ideal time to spray the nest is at night when they're all home. Unfortunately night time is the hardest time to spray because you can't see anything, so dusk or dawn may have to do. Otherwise, you can spray them during the day and expect to see bees returning to the nest all day and swarming around the door to their nest wondering what the heck happened to their home. When they realize their home and queen have been destroyed they probably won't attack you in a mad frenzy, as they're most defensive when defending their home and queen. Once the queen is dead the bees don't seem to know what to do. However, they may still sting, so watch out.

The yellow jacket nest starts out in the spring with only one bee (the queen). She builds a small paper nest about the size of a golf ball, then starts laying and hatching eggs. By fall, the nest will be the size of a football and have thousands of bees in it. Obviously, a good time to destroy a nest is in the spring when it is tiny. The longer the nest is allowed to expand, the more bees it will have and the more effort it will take to get rid of it. Of course, the bigger a hidden nest gets, the easier it is to locate because there are more bees flying in and out of it. Fall is the worst time for roofers as far as bees are concerned, and in September and October the roof worker needs to be especially vigilant.

Do not try to kill yellow jackets or any other bees with general insect spray - it doesn't work and it just makes the bees mad. Use only wasp and hornet spray. Finally, if you have to approach a nest closely to spray it, wear coveralls, gloves, and a bee veil. I always keep these essential materials in my work truck and they have saved me from bee stings many a time. If the nest is in the ground you can pour a flammable liquid down the hole and light it (unless the hole is near a house or other object that you don't want to burn to the ground).

Honeybees aren't as mean as yellow jackets, but will defend their nest just as violently if need be. Their nests are rarely exposed and are made of wax cells where the bees store honey and hatch out young. You can locate their nests in the same manner as yellow jacket nests - by looking for their flyway. You will see them flying in and out of a hole in the building.

Again, their behavior is affected by the weather and they don't want you anywhere near their nest when rain is impending (rising humidity), probably because their nests are vulnerable to rain. They too are friendliest during sunny, dry weather. Honeybees (and yellow jackets) tend to attack black and rough surfaces first. That's why they go for your hair. Maybe they evolved in such a manner that they think everybody is a bear trying to steal honey and so they dive for that part of you that looks like a bear. Knowing this, you would never approach a bee nest wearing dark, rough clothing. Beekeepers wear smooth, white coveralls. Honeybees generally will not go out of their way to sting you like yellow jackets will, but if the weather's bad and you resemble a bear - look out!

I have never destroyed a honeybee nest and probably never will - they're beneficial insects. It's easier to wear protective clothing around them than it is to try to kill them. A full grown honeybee nest will contain tens of thousands of bees that will pollinate many flowers as well as produce much honey and beeswax. But they do live in the walls of old houses where they remain year after year, although they're not as common as yellow jackets. The solution to honeybees is to avoid them. If you can't, then wear protective clothing and work around them when they're in a good mood.

Bald-faced (white-faced) hornets are mean like yellow jackets. They live in paper nests which can get rather large (football size or bigger) and hang down from the eaves of a house, containing as many as 10,000 hornets. I have seen their nests hanging on almost anything, including tree branches and brush, and saw one nest sandwiched between two windows thereby exposing the entire inner workings of the hive. They are nasty critters that can't be trusted. They'll come after you aggressively and *should* be destroyed.

Paper wasps are not mean. They are a relatively benign stinging insect. They *will* sting you though, if you mess with their nest. You can get quite close to a paper wasp nest and they won't come after you unless you vibrate the nest by banging near it. I do spray their nests and I do recommend it when you have to work near one because I've seen a number of roofers get stung by these fellows.

One of my workers was stung by a paper wasp while tying a ladder to a rain gutter. The nest was up under the slate at the eaves and he stuck his hand practically right in it, then developed an allergic reaction when stung. One minute he was normal and the next he was covered with hives, his face was swollen, and he was lying on the tailgate of the truck. I watched another fellow also get stung in the hand by paper wasps for the same reason - reaching under the eaves of a dormer (a favorite spot for paper wasps) without looking first. Another roofer I know fell off the peak of a roof after exposing a paper wasp nest while removing ridge iron. The wasps came flying out at him and he backed off the roof in panic, falling nearly thirty feet. He was lucky - he only broke his leg! You must *always* look under eaves and in any area that may hide a nest of bees before getting your face (or hands) in it. Remember to find them before they find you! Paper wasps love to build nests under ridge iron, so always remove ridge iron carefully (and have a can of hornet spray handy in the summer months) expecting live wasp nests to be underneath (we'll mention this again when discussing ridge iron replacement).

Bumble bees, and carpenter bees look very similar. They're round, black and yellow and fuzzy, and they're usually not mean. Bumble bees live in colonies often in the ground, while carpenter bees live in holes they bore in wood. There are large carpenter bees and small ones. The small ones seem to live in groups, while the large ones are loners. You'll often see the large carpenter bees when on the roof of old buildings because they live there too. They're very territorial and will hover in the air like a tiny helicopter, staring at you and trying to intimidate you. If any other insect happens to fly past at that moment, the carpenter bee will chase after it until it leaves its territory, then it will come back to you as if to ask "What are *you* still doing here?!" I've never known anyone to be stung by one, although I've had to swat a few with my slate ripper just to get rid of the persistent little devils. Their danger lies in their ability to startle you, which you can't afford when working at heights.

Finally, **mud daubers** are common on old roofs. They're long, thin, purplish wasps that nest in clumps of mud they build on (or in) chimneys, under flashings, and especially in attics. They're not mean and are usually not a problem.

BATS

Bats come in various sizes and shapes and are very common around, in, and under old slate roofs. They are not a danger except that they may startle you if they fly out at you, and your instinctive recoil can knock you off the roof. However, if you're aware of their presence and expect them, you won't be startled if one does appear.

Bats love to hang out under ridge iron, under chimney flashing, in old chimneys (sometimes between the bricks), under loose slate, and especially in attics. They can get in to the roof through almost any hole, and are almost impossible to keep out. Again, these are beneficial creatures that eat a lot of bugs, so the solution to bats is not to kill them but to learn to live with them. I've seen people try getting rid of bats by putting mothballs in their attics (not very effective), and by using electronic pest repellent (might work). Some people swear that you only need to keep a light on in the attic to keep them away, but this is hearsay and I can't vouch for it.

Bats hibernate in the winter, then become active again when it warms up. In the fall when the days are nippy, bats, like wasps, move in slow motion and are easily captured. They do have a lot of teeth and might bite, so I don't recommend handling them without gloves. Of course, everyone says they carry rabies, but this threat may be exaggerated.

You can usually locate bats easily because first, their droppings give off a characteristic odor, and secondly, the bats themselves give off a characteristic high-pitched squeak or chirp when agitated. Furthermore, you can hear them scrambling and scratching under ridge iron when you've climbed on the roof and disturbed them. On one roof I shoved my slate ripper under a broken slate to remove it, and when I pulled the ripper out a live bat came out with it! If you have an old house with a slate roof, you probably have bats. If you expect them and look for them, they won't startle you and cause an accident.

CHIMNEYS

Never grab onto a chimney for support unless you're quite sure that it is solid (most aren't). Many of the old chimneys on slate roofs have soft mortar, and the bricks can easily be lifted apart. This becomes a hazard when someone grabs onto a chimney to steady himself when moving along a ridge, and the bricks give way. Also, when taking chimneys apart, remember that bats do live in them, and sometimes bees do too. Find them before they find you. Finally, chimney swifts (birds) also live inside chimneys in shelf-like nests fastened to the side of the chimney. When you peer down into the top of the chimney, a startled bird may fly out in your face. Be prepared.

WIND AND LIGHTNING (AND HEAT AND RAIN)

Wind is an obvious hazard when working on roofs. It's much windier on top of a roof than on the ground, and on very windy days it may be a good idea to stay off the roof. The danger of wind is magnified when you are carrying a hook ladder up onto a roof, across the roof, or back down off the roof. The wind will catch the ladder and push it, which may cause you to lose your balance. Also, wind will readily knock a ground ladder off a house, so *ground ladders must always be tied to the house* to prevent this from happening. Aluminum ladders are light and easily blown

SUN   LIGHTNING   RAIN   WIND   ICE

DANGER

SLATE ROOFS ARE SLIPPERY WHEN WET

WEATHER HAZARDS ON SLATE ROOFS

down onto a person, car, or other property (not to mention the damage to the ladder, or the embarrassment to the stranded roofer).

Roofers are human **lightning** rods. *Always* vacate the roof as soon as any lightning becomes evident. Thunderstorms have a way of coming out of nowhere, so this can be a real hazard, especially when aluminum ladders are being used, as aluminum is a great conductor of electricity. Don't stand on the ground near a standing aluminum ladder during an electrical storm either.

Heat can be terrible on a slate roof during the summer months. I have personally measured a temperature of 135 degrees Fahrenheit on the working surface of a slate roof, and have read that 140^0 is not uncommon. People do keel over from heat stroke, and if you are susceptible to this sort of thing stay off a slate roof during the heat of a summer day. Furthermore, sunburn can lead to skin cancer in later years, so protect yourself from excessive, chronic solar exposure. **Rain,** dew, and of course frost make the surface of slate roofs wet and slippery. As a rule, *never work on wet or frozen slate roofs.*

ELECTRICAL WIRES

ELECTRIC LINES ARE A MAJOR HAZARD WHEN WORKING WITH ALUMINUM LADDERS. ALWAYS LOOK FOR THEM BEFORE PUTTING UP OR TAKING DOWN A LADDER.

Electrical wires are a very serious hazard. If your aluminum ladder happens to hit a high-voltage wire while you're touching the ladder, you're dead - instantly. Electric lines may contain as much as 7200 volts, or may be stepped down to 2400 volts until the lines reach the transformers located on a pole near the house, where the voltage is reduced to 110/220 volts. The "high voltage" lines are certain death, and the 110/220 lines can also kill, so *always* be aware of the location of electrical lines and keep your aluminum ladders away from them.

When electrical lines enter the house from a public electric service source there will always be at least two lines, and they will attach to the house on an insulated mount, usually leading to an electrical meter. This differs from telephone lines because telephones only need a single line (although if the house has several phones there may be several phone lines), and phone lines usually don't lead to an insulated mount, and never to an electrical meter. If you see lines entering the building you want to work on and aren't sure what they are, avoid them.

Electricity causes muscles to contract. If you're holding on to a ladder and you hit an electrical line with it, the electricity will cause your hand muscles to contract forcing you to remain holding on to the ladder. If you have to touch a surface to see if it is conducting electricity (don't try this), touch it with the *back* of your hand, then, if the surface is "hot" it will contract your muscles and throw your hand away from the surface. Better yet, if you have any doubts about whether a metal surface on the outside of a house is "hot" or not (it happens), call the electric company and let the experts take care of it.

I have heard several accounts of people getting electrocuted while working on construction jobs. In one case, a man was carrying sheet metal roofing up onto a barn roof and the wind caught it and blew it against a high voltage line while the guy still had hold of it (he died instantly). One highly publicized case involved three men trying to take down a metal flag pole in the front yard of a town hall. The pole hit a high voltage line and the men stood in the yard in the center of town welded to the flag pole literally frying to death in front of a crowd of horrified onlookers for a full half hour until the electric company could shut off the power (the three men all died instantly).

Interestingly, when I contacted my local electric company (West Penn Power) to get accurate information on voltage levels of power lines, they routed my call to four different people (receptionist, engineer, public relations officer, and safety manager) before finally stating that they wouldn't give me that information. They told me to look it up at the library. So much for the "experts." Go figure.

CO-WORKERS

Co-workers can be an unintentional hazard. Like the un-informed wife who took off in the car without first ensuring that her actions wouldn't adversely affect her husband's safety, co-workers cannot always be relied on to maintain safe working conditions. As stated earlier, don't rely on the judgement of co-workers when your own safety is at stake. Co-workers can drop things on you from above, throw or drop things on the ground without looking, bump into you carelessly on the roof causing you to lose your balance, knock you

Even co-workers can create hazards. Watch out!

ROOF CEMENT

while carrying a ladder on the roof, put up a ladder recklessly, attach a ladder hook to a ladder carelessly, install a roof bracket incorrectly, fail to tie ladders in place securely, overload roof scaffolds with too much weight, etc. This is not to suggest that co-workers are bad, but accidents can and do happen, and you must keep an eye on your co-workers when working at heights as if they were hazards themselves, because they can be. Conversely, co-workers can greatly enhance your safety, but accidents occur when people get tired, cranky, impatient, over-worked, and careless. So consider this "a word to the wise."

3) USE THE PROPER EQUIPMENT

You can safely do almost any job on a slate roof with the following equipment: **ground ladders, roof ladders** (also called hook ladders), **roof jacks** (also called roof brackets), **planks,** and **ladder jacks** (also called ladder brackets). Sources of tools and equipment are listed in the next chapter, although ladders, roof jacks, ladder jacks, and planks are available at almost any lumber yard. The only specialty tool is the ridge hook (used to make a hook ladder). In severe cases a **"harness"** may be necessary, which is a safety belt that wraps around the worker and ties on to something to prevent a fall. In 30 years of working at heights, I have had to wear a harness on only one job, and I generally avoid anything with ropes on a slate roof, because ropes impede mobility and create a tripping and snagging hazard. People who are not so accustomed to working at heights may want to use a harness more often. Some people use them all the time, and in some cases they're required by government safety codes. Unfortunately, government safety codes are sometimes based upon theory and are dreamed up by well-intentioned bureaucrats sitting in offices, rather than by the people who actually do the work. In which case, instead of being helpful they can create incredible burdens upon the workers, if not safety hazards themselves (like ropes to trip over).

The tools mentioned above will be discussed in greater detail in the next chapter (Tools and Equipment), but let's look at the safety element first.

GROUND LADDERS

A ground ladder is one that stands on the ground and is used to get up to the roof (see Figure 10.1). This is the typical ladder that most people are familiar with. Aluminum ladders are recommended because they're light in weight and easy to handle, although aluminum ladders do conduct electricity and are very dangerous around electrical wires. Ladders must be put up properly or they can be very hazardous. Ladders will blow off the house if not tied on, and are typically tied to a rain gutter, or a nail or hook driven into the fascia. They must not be set up too steeply, or not steep enough (at too great an angle), and must be put up plumb (perpendicular to the force of gravity - not leaning sideways). Although smaller ladders can be put up by one person, longer ladders should be put up by two people, for safety's sake.

HOOK LADDERS AND RIDGE HOOKS

Once you get up to the eaves of a roof, you generally must use a hook ladder to get to the peak. Hook ladders enable the roofer to gain access to any spot on a roof safely, and without damaging the slate (Figure 10.3). Don't try to climb up a slate roof by walking on the slate, hanging on a rope. Not only does this damage the slate, but a loose slate can slip out when you step on it and you will suddenly know what it's like to ride a skateboard forty feet in the air.

Good ladder hooks (ridge hooks) are inexpensive and will attach to almost any ladder. Typically, you would sepa-

Figure 10.3

HOOK LADDERS ARE SAFER THAN EITHER ROPES OR "CHICKEN LADDERS"

Swiveling angle-iron end piece

Ladder Hook (also called ridge hook) shown attached to ladder. Hook ladders allow for relatively safe access to the top of roofs. This type of hook is best suited for slate roofs because it has a hinged, flat piece of metal contacting the slate, which prevents breakage. This hook is easily removable and can be attached to any pair of rungs on the ladder, allowing for the working length of the ladder to be adjusted. The manufacturer of this hook specifies that they be used in pairs (two hooks on the end of a ladder instead of one), but two hooks on one ladder are sometimes heavy and hard to handle, thereby posing a hazard themselves, so roofers *typically* use only one per ladder.

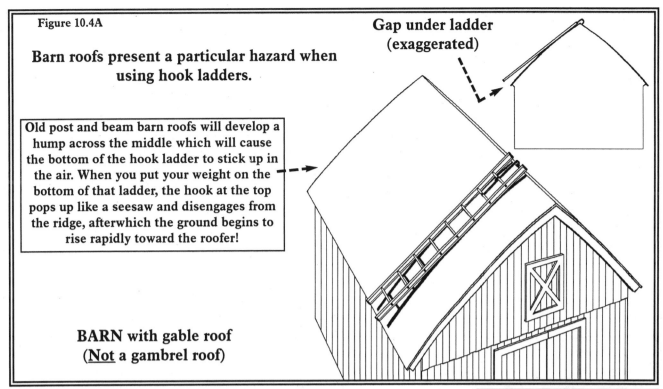

Figure 10.4A

Barn roofs present a particular hazard when using hook ladders.

Gap under ladder (exaggerated)

Old post and beam barn roofs will develop a hump across the middle which will cause the bottom of the hook ladder to stick up in the air. When you put your weight on the bottom of that ladder, the hook at the top pops up like a seesaw and disengages from the ridge, afterwhich the ground begins to rise rapidly toward the roofer!

BARN with gable roof (Not a gambrel roof)

rate an aluminum extension ladder into its two halves, then attach a hook to one of the halves in order to make a hook ladder. The ladder is simply slid up the roof until the hook drops over the peak and secures itself. Usually, the hook is facing upward when the ladder is slid up the roof, then the ladder is flipped over when the hook reaches the top. Some hooks have little wheels attached to make it easier to slide them up the roof.

The hook ladder must be long enough to reach from the peak of the roof to the eaves in order to be able to climb onto it from the ground ladder. It is critically important that the ridge hook be secured tightly to the hook ladder, as several contractors have told me that their ridge hooks have come off at the most inopportune times. This is somewhat surprising to me, as I have used hook ladders since 1968 and have never had a hook come off a ladder. But then, I do keep a safety attitude, and check my equipment regularly.

A greater danger when using hook ladders arises on old barns, and some old buildings. Such roofs tend to have a hump in the middle, halfway between the peak and eaves (Figure 10.4A). This causes the bottom of the hook ladder to be lifted off the roof several inches, and if you put your weight on the bottom of the ladder, the top will pop up (the see-saw effect), the hook will disengage, and you and your ladder will quickly plummet earthward. *This is a very serious hazard when using hook ladders on old slate roofs, and you must be very careful that the bottom of the hook ladder is lying tightly against the roof when the top is firmly secured.* Otherwise, a block of some kind (a piece of 2x4 or 4x4) must be tied to the bottom and underneath the hook ladder to prevent the seesaw effect from ruin-

Figure 10.4B
This is a common type of hook ladder <u>not</u> recommended for use on slate roofs, because the hooks are pointed and will damage the slate.

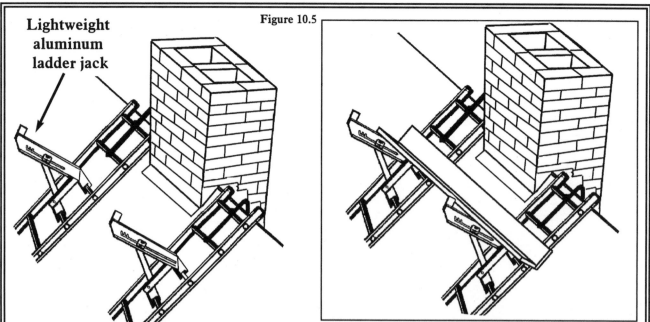

Figure 10.5

Lightweight aluminum ladder jack

Two hook ladders with ladder jacks are positioned on either side of a chimney (left). 2"x10" planks are laid across both the ladders and the jacks to make a safe and easily installed roof scaffold (right). Double-wide planks can be laid on both the ladder and the jacks to make a platform with greater capacity. If this scaffold is expected to bear heavy loads, then use two ridge hooks per ladder. Access the other side of the chimney from roof jacks and plank, or use two more hook ladders with longer planks Be careful not to overload the roof scaffold, as too much weight can mean disaster.

ing your life. Also, a co-worker can sit on the top of the hook ladder to make sure it doesn't disengage.

Hook ladders can also be hazardous when they have to be carried around on a roof, especially long hook ladders. Typically, the hook ladder is carried along the ridge like a tight-rope walker carries his balancing pole along a tight rope. This is the easiest way to move a ladder around on a roof, but it requires a good bit of balance. However, a ladder will catch the wind much more readily than a pole will, and a gust of wind can knock a person off balance when carrying a hook ladder along a ridge. Alternatively, the person can sit on the ridge and drag the hook ladder along the roof, but this actually requires a lot more effort (but little balance).

Many contractors will use what are called "chicken ladders" instead of hook ladders. These are home-made contraptions, usually made of wood, that function in the same manner as hook ladders. They are not as safe or convenient as an aluminum ladder with a ridge hook attached to it though, and therefore are *not* recommended.

LADDER JACKS, ROOF JACKS AND PLANKS

Again, these tools are discussed in greater detail in the next chapter, but are mentioned here because you cannot safely work on slate roofs, *at times*, without them. Hook ladders only gain access to roofs that have horizontal ridges, and not all slate roofs have them. Some come to a point, some have flat tops, while others, such as gambrel roofs (see Figure 10.2), do have horizontal ridges but they can't be easily reached with a hook ladder. When you can't get a hook ladder up to a ridge, then you have to use roof jacks and planks in order to get on the roof.

The terms roof *jacks* and ladder *jacks* are old-fashioned roofers terms. Manufacturers call these things roof brackets and ladder brackets. They will generally be referred to as roof and lad-

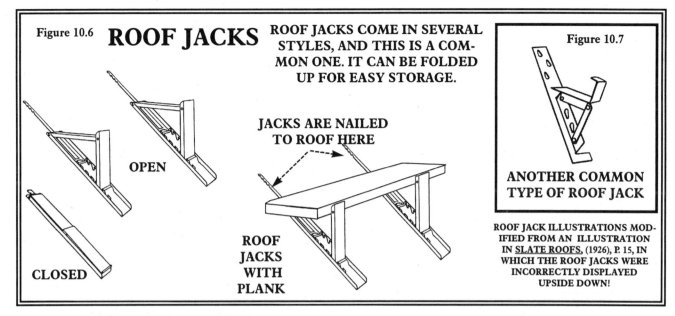

Figure 10.6

ROOF JACKS

ROOF JACKS COME IN SEVERAL STYLES, AND THIS IS A COMMON ONE. IT CAN BE FOLDED UP FOR EASY STORAGE.

OPEN

CLOSED

JACKS ARE NAILED TO ROOF HERE

ROOF JACKS WITH PLANK

Figure 10.7

ANOTHER COMMON TYPE OF ROOF JACK

ROOF JACK ILLUSTRATIONS MODIFIED FROM AN ILLUSTRATION IN SLATE ROOFS, (1926), P. 15, IN WHICH THE ROOF JACKS WERE INCORRECTLY DISPLAYED UPSIDE DOWN!

der jacks in this book for tradition's sake.

Roof jacks nail to the surface of the roof, *right through and on top of the slate*, and support horizontal planks which are used as working platforms (see Figures 10.6 and 10.7). A ladder can be carefully set on these planks and laid flat on the roof to get up higher. Roof jacks, when properly nailed in place and used with a good quality, proper size plank, are quite safe and indispensable for slate roof work. They can be nailed into a slate roof and removed from the roof without damaging the roof (this procedure is described in the next chapter).

A *ladder jack* does not attach to the roof, but instead attaches to a ladder (see Figure 10.5). They must be used in pairs, much like roof jacks, and they have the same purpose as roof jacks, namely to support planks so as to provide a work platform. They come in handy on slate roofs when two hook ladders are positioned on the roof next to each other, and a ladder jack is attached to each one, then one or more planks laid across the ladder jacks. This allows for a solid roof platform to be quickly and easily constructed without the need to nail anything into the roof. This "roof scaffold" system is particularly useful when working on chimneys, as a hook ladder can often be positioned on either side of the chimney, and planks can be run across from one ladder to the other just below the chimney (more on this in Chapter 15).

Now let's take another look at the guy who tied himself to his car and got yanked off the roof. He obviously had a ground ladder or he wouldn't have been on his roof at all. If he had had a hook ladder he would have been able to climb onto his roof without the need for a rope. Two hook ladders would have enabled him to get practically anywhere on his roof (one to go up and down from the ground ladder, and one to move around the roof by walking along the ridge - the "sidewalk" on the roof). If he had the sort of roof where a hook ladder wouldn't work, he could have nailed a pair of roof jacks to his roof (working from his ground ladder), laid a plank across them, and climbed onto this roof platform. He could have then set another ladder onto this platform to get to the spot that needed fixed. If he still couldn't reach the spot, he'd have to install more roof jacks and planks until he got where he wanted. Then if his wife got a sudden urge to run to the store, he'd survive unscathed.

Bats were rudely evicted, live wasp nests were destroyed, and feet were baked - all in a summer morning's work on a slate roof. At 95 degrees Fahrenheit in the shade, roof soot clings to sweat as evidenced by three slaters taking a merciful lunch break.

Chapter Eleven
Basic Tools and Equipment

D on't read this chapter unless you've read Chapter 10 (Safety First). Chapter 10 introduces you to some of the equipment required for slate work and explains some of the hazards one can, and *will* run into when working on slate roofs. Chapter 11 will show you how that equipment, as well as hand tools, is meant to be used.

Slate work is a specialized endeavor requiring some unusual tools and equipment not readily available at a hardware store. If you don't use the right tools, you'll have a very hard time doing any work on a slate roof. The tools you will need *are* available, however, and it is strongly urged that if you intend to do any work on any slate roof, you invest in a few tools appropriate for the job. Slate tools are not prohibitively expensive and are well worth the investment if they enable you to keep a slate roof alive. (See sources of tools at the end of this chapter.)

SLATE RIPPER

There are several hand tools necessary for working on slate roofs, and the most important is the **slate ripper**. The ripper is a long, sword-like steel tool that is slid under a slate to hook the (usually two) nails holding the slate, and pull or rip them out. Once the nails are pulled out the slate can be removed. This procedure is necessary when replacing slates, as the ripper enables the worker to remove the old, broken slate, which must be done before a new slate can be slid into place. Slates are also removed from existing slate roofs when flashing metal is replaced, or when holes are cut into the roofs to install (for example) skylights. The ripper is as important to the slater as the hammer is to the carpenter.

Although slate roofs are unique in many ways, they have one quality that separates them from most other roofs: they can be taken apart and put back together. A slate roof should be seen

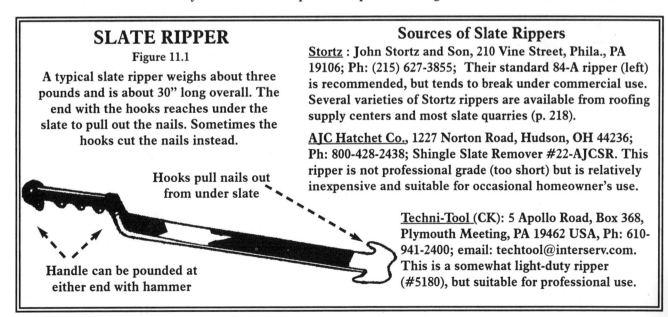

SLATE RIPPER

Figure 11.1

A typical slate ripper weighs about three pounds and is about 30" long overall. The end with the hooks reaches under the slate to pull out the nails. Sometimes the hooks cut the nails instead.

Hooks pull nails out from under slate

Handle can be pounded at either end with hammer

Sources of Slate Rippers

<u>Stortz</u> : John Stortz and Son, 210 Vine Street, Phila., PA 19106; Ph: (215) 627-3855; Their standard 84-A ripper (left) is recommended, but tends to break under commercial use. Several varieties of Stortz rippers are available from roofing supply centers and most slate quarries (p. 218).

<u>AJC Hatchet Co.</u>, 1227 Norton Road, Hudson, OH 44236; Ph: 800-428-2438; Shingle Slate Remover #22-AJCSR. This ripper is not professional grade (too short) but is relatively inexpensive and suitable for occasional homeowner's use.

<u>Techni-Tool</u> (CK): 5 Apollo Road, Box 368, Plymouth Meeting, PA 19462 USA, Ph: 610-941-2400; email: techtool@interserv.com. This is a somewhat light-duty ripper (#5180), but suitable for professional use.

as an entity consisting of thousands of parts, each removable and replaceable. Those parts are the slates themselves, and they're removable with the help of a slate ripper.

Like an automobile, slate roofs need to be "tuned up" now and then. A car needs its spark plugs replaced, while a slate roof needs a few slates replaced occasionally. When you consider that an average slate roof may have 3,000 slates on it, and that many slate roofs after a hundred years have only had maybe 30 slates replaced, then you can see that 99% of the roof has *not* needed repaired in a century's time. Yet, many people will convince themselves that a 1% breakdown after a century means it's time to replace the roof. This is similar to saying that when the spark plugs of your car wear out it's time to replace the car!

The reason this may be hard to understand is because slate roofs follow a time span that is not of a human scale. Humans typically last maybe 80 years, while slate roofs may last hundreds. So when people see an 80 year old roof they jump to the conclusion that its time is up. This misperception is compounded by the fact that America is a land of the throw-away culture, where people take it for granted that when something breaks, it is simply thrown away and replaced. Slate roofs, however, don't break. Like cars, they wear out, but unlike cars it's usually over a period of time which may be centuries, depending on the type of slate. And like cars, they can not only be tuned up, but also completely overhauled.

A standard overhaul of a slate roof requires replacement of such things as the metal flashings, chimney tops, broken slates, and possibly some roof sheathing (wood). A total and drastic overhaul involves removing the entire slate roof, replacing the wood underneath, then re-nailing the same slate to the new wood with new nails. At this point in time (end of the Twentieth Century), many slate roofs do need overhauled somewhat if they're going to last another century. However, like a mechanic who can't get your spark plugs out without a spark plug wrench, a slater can't take your roof apart without a slate ripper.

Like any craftperson's tool, the ripper is a tool that becomes more versatile as the user gains more experience. It can pull nails out, push nails out, and cut nails by both pushing and pulling. It can be used to feel around under slate to find obstructions, old repairs, hidden nails under tarred slate, etc. It even makes for a suitable bee swatter in an emergency. It should never be used to pry with, however, as it bends easily and is not designed for prying.

A ripper sometimes breaks when beating on it with a hammer, especially in cold weather (I've broken many), and usually they just snap in two somewhere along the shaft, although sometimes a hook will break off or the handle will snap in two. You're *supposed* to beat on the ripper with a hammer, but that doesn't mean it can't eventually break. Rippers are designed to pull out roofing nails and not some of the larger nails (such as 8 penny and 16 penny nails) that some roofers will incorrectly use to nail replacement slates. A good way to break a ripper is to hook it onto an eight penny nail and try to pull it out by feverishly beating on the tool. Broken rippers can be welded back together, however, and used again for quite some time.

Sometimes a ripper will bend the nail but not pull it out, and even though the old slate will come out, the old bent nail remains behind and blocks the replacement slate from sliding in. This is a common pain in the you-know-what, and the only solution is to use the ripper to bend the nail back and forth until it breaks off. That's one reason the point of the ripper is slightly indented - so nails can be pushed with the tool as well as pulled. Sometimes there is absolutely no way to get a ripper hooked on a nail, and then the nail must be cut by using the point of the ripper and pounding on the end of the handle as in Figure 11.3. This is only recommended as a last resort because you risk pushing the nail up under the slate where it won't come out at all (without removing more slates).

The pointed end of the ripper can also make a suitable chisel for scraping old roofing

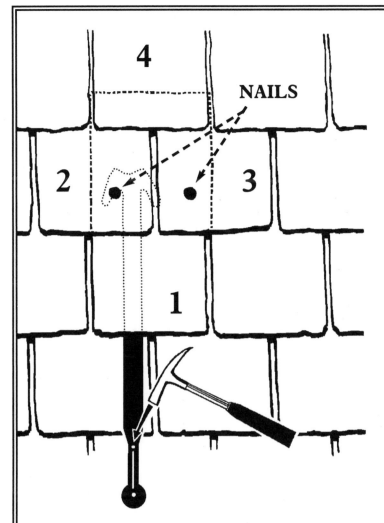

NAILS

Figure 11.2

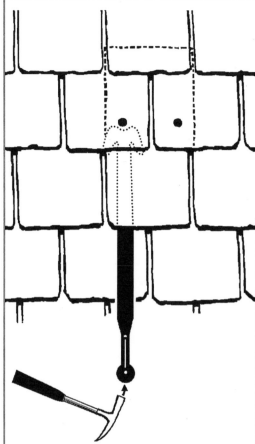

Figure 11.3
A good hard rap or two on the nail with the tip of the ripper will loosen it up and make it a lot easier to get the ripper hooked on the nail so it can be pulled out. Often the slates are too tight to get a ripper under far enough to hook the nail, and this little trick can save a lot of grief.

USING THE SLATE RIPPER

The diagram above shows the location of the nails holding slate #1 in place (slates 2, 3 and 4 are overlapping slate #1). In order to remove slate #1 from the roof, the two nails must be pulled out, one at a time, with the ripper, as shown. The ripper is slid underneath the slate to be removed. A hammer is used to pound downward on the handle of the ripper in order to force the nails out, while the worker pulls on the ripper with the other hand. Once the slate is out, a new slate can be installed (see Chapter 13). Sometimes it is difficult to get the ripper under the slate and hooked on the nail, because the slates are too "tight". This problem can be alleviated by shoving a couple of nails under the bottom edge of the slate to wedge it up a bit, and/or by pounding on the slate nail with the tip of the ripper before trying to remove it (see Figure 11.3).

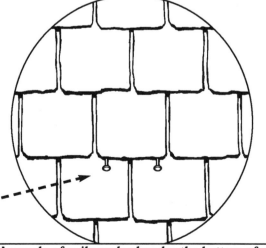

A couple of nails wedged under the bottom of a slate may help to make room for a ripper.

cement off the surface of a slate. I've seen many tarred slates cleaned up this way only to reveal that there was no reason whatsoever for the slate to be tarred in the first place- no hole, no crack, no exposed nail (evidence that Tarbaby Teddy exists!).

Hand tools can get very hot on slate roofs during the summer months, and rippers are particularly notorious for this. They actually get too hot to handle and gloves must be worn when using them. When the ripper is not being used, the point is slid far enough under a slate on the roof so that the ripper will remain safely lodged, out of the way, and easily accessible until it is needed.

SLATE CUTTER

Contrary to popular opinion, roof slates are easy to cut using a slate cutter, a simple hand tool which is similar in principle to a paper cutter. The slate cutter has a long arm with a cutting blade on it; the slate is worked through the blade with one hand while the cutting arm is operated with the other (see Figure 11.4).

No electric tools are needed to cut slate, or to work with slate in any way. Remember that slate roofing became an art long before electricity came into common usage. On the other hand, some slaters will use a masonry blade on a circular saw to cut some of the very hard, *new* slate, when delicate cuts are required. Generally speaking though, old, recycled slate such as used for restoration work need nothing but a hand-held manual slate cutter. The disadvantage to using a

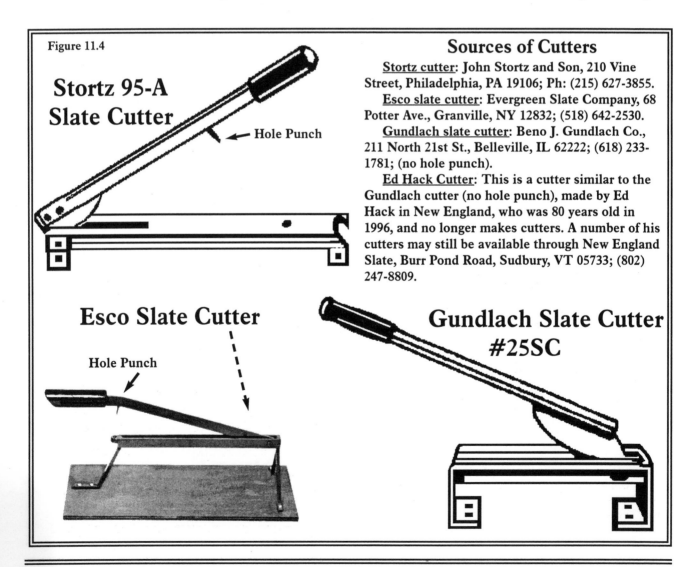

Figure 11.4

Stortz 95-A Slate Cutter

← Hole Punch

Esco Slate Cutter

Hole Punch

Gundlach Slate Cutter #25SC

Sources of Cutters

Stortz cutter: John Stortz and Son, 210 Vine Street, Philadelphia, PA 19106; Ph: (215) 627-3855.

Esco slate cutter: Evergreen Slate Company, 68 Potter Ave., Granville, NY 12832; (518) 642-2530.

Gundlach slate cutter: Beno J. Gundlach Co., 211 North 21st St., Belleville, IL 62222; (618) 233-1781; (no hole punch).

Ed Hack Cutter: This is a cutter similar to the Gundlach cutter (no hole punch), made by Ed Hack in New England, who was 80 years old in 1996, and no longer makes cutters. A number of his cutters may still be available through New England Slate, Burr Pond Road, Sudbury, VT 05733; (802) 247-8809.

masonry blade on a circular saw to cut slate is that the saw cut leaves a square edge, whereas all roof slates have beveled edges. The hand-held slate cutters leave a beveled edge on the slate, and this matches the original slate much better than a cut from an electric saw.

A good slate cutter can cut curves in slate both convex and concave, and punch holes in the slate, as well as make a simple straight cut. Some commercial slate cutters today don't have a hole punch on them and can't cut concave curves. Concave curved cuts are necessary when the slater is making a decorative cut in a square slate to match the curved design on an existing roof (see Figure 12.9, page 200). The cutter of choice has often been the O. Pearson Manufacturing Company Asbestos Shingle Cutter, which makes a great slate cutter, and has a hole punch. I've used the same Pearson cutter for 30 years and it still cuts as well as the day I first used it, never needing sharpened or any blade replacement (although the hole punch wore out years ago). I inherited the cutter from my mentor, Pete Odrey, of Lyndora, PA, who used it for decades before I got it. Unfortunately, Pearson went out of business a number of years ago and their cutters are no longer available. However, a cutter modeled after the Pearson cutter, the *Stortz Cutter*, is being manufactured and is available, as are several other cutters (see Figure 11.4 as well as sources of tools at end of this chapter).

Old slate tools can also be found at flea markets, auctions, garage sales, and antique stores. Old tools can be superior to new ones, and are often preferred by professional slaters. So you may still find one of the old Pearson cutters collecting dust in someone's basement or attic, as well as an old Belden, Pecto or Red Devil ripper or slater's hammer.

A slater will use the cutter while on the roof, so it must not be a cumbersome or awkward tool. Most cutters are designed so that they can be mounted on a board, which is fine when using the cutter on a flat surface. But when up on a roof, a board is not necessary as the slate is being cut by a roofer usually crouched in an awkward position, and the bottom half of the cutter is simply tucked into the roofer's tool belt as the slate is cut. The cutter is then opened up and laid over the ridge of the roof for safe storage, or propped in a rung of a hook ladder until needed again.

A nifty little slate cutter is the "hand-held" cutter, which is small (about 10" long) and can be carried in a tool belt. Made in France and sold by Stortz, it will cut roof slate much as a pair of tin snips will cut metal (Figure 11.4B).

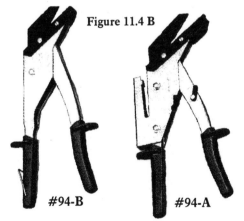

Figure 11.4 B

#94-B   #94-A

HAND HELD SLATE CUTTERS
Measuring 10" long, these handy little cutters will go through a piece of standard thickness roof slate, and still fit in a tool belt. The one at right also punches holes in slate. Made in France and sold by Stortz.

SLATE HAMMER

A slate hammer is a unique tool that comes in very handy for a slater because it not only pounds nails, but also pokes holes in slate, and some slate hammers can be used to cut slate. It's not necessary for a homeowner to have a slate hammer, however, because a regular 16 ounce carpenter's hammer does a fine job of pounding nails and a good slate *cutter* does a better job of cutting slate than a slate hammer. The hammer is most handy when used to poke nail holes in slate, although a cutter with a hole punch will do the same job, and a carpenter's hammer and a nail will poke a hole in slate too. If you intend to do any significant amount of work on slate roofs, however, a slate hammer is a highly recommended tool.

New slate hammers, manufactured in Germany and

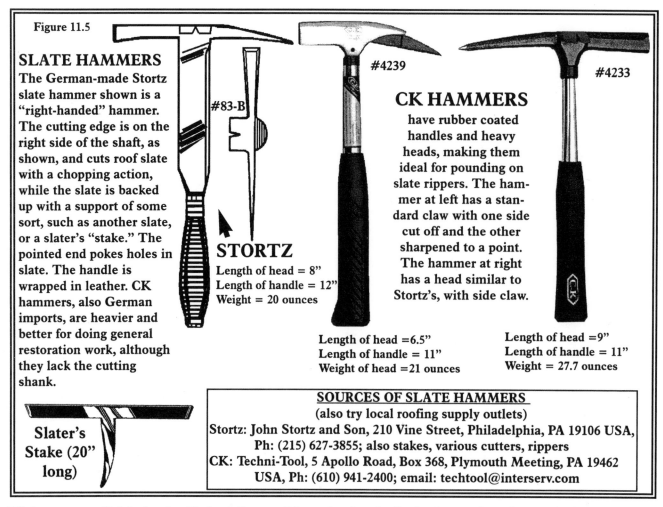

Figure 11.5

SLATE HAMMERS
The German-made Stortz slate hammer shown is a "right-handed" hammer. The cutting edge is on the right side of the shaft, as shown, and cuts roof slate with a chopping action, while the slate is backed up with a support of some sort, such as another slate, or a slater's "stake." The pointed end pokes holes in slate. The handle is wrapped in leather. CK hammers, also German imports, are heavier and better for doing general restoration work, although they lack the cutting shank.

Slater's Stake (20" long)

#83-B

STORTZ
Length of head = 8"
Length of handle = 12"
Weight = 20 ounces

#4239

CK HAMMERS
have rubber coated handles and heavy heads, making them ideal for pounding on slate rippers. The hammer at left has a standard claw with one side cut off and the other sharpened to a point. The hammer at right has a head similar to Stortz's, with side claw.

Length of head =6.5"
Length of handle = 11"
Weight of head =21 ounces

#4233

Length of head =9"
Length of handle = 11"
Weight = 27.7 ounces

SOURCES OF SLATE HAMMERS
(also try local roofing supply outlets)
Stortz: John Stortz and Son, 210 Vine Street, Philadelphia, PA 19106 USA,
Ph: (215) 627-3855; also stakes, various cutters, rippers
CK: Techni-Tool, 5 Apollo Road, Box 368, Plymouth Meeting, PA 19462
USA, Ph: (610) 941-2400; email: techtool@interserv.com

Wales, are available in the United States. The selection is limited to only a few varieties of hammers: a right hand and a left hand Stortz hammer (depending on which side of the shaft the cutting edge is on); plus two varieties of CK slater's hammers, which don't have cutting edges. Older Belden and Pecto hammers can still be occasionally found at flea markets and antique stores, many of which have cutting edges on both sides of the shaft and are useful for both right and left handed people. The older flea-market hammers are often preferred by professional slaters.

The author's hammers of choice for restoration work are the CK hammers (either one), which are excellent hammers because they have enough weight to effectively pound on a ripper and to pound in the larger nails used on roof jacks, roof sheathing etc. The Stortz hammer is too light for restoration work, although it is excellent for installing slate as it has enough weight to pound roofing nails. Stortz was in the process of developing a heavier slater's hammer at the time this book was written, so by the time you read this, they may have a hammer better suited for the rigors of restoration work.

Incidentally, the author had to go to Wales to acquire his CK hammers, which heretofore had never been available in the United States. The manufacturer, however, has agreed to make these hammers available in the US via one of its distributors, Techni-Tool (see Figure 11.5).

A common way to cut slate when no slate cutter is available is to use a slate hammer with a cutting shank (Stortz hammer for example) and slater's **stake**. The stake is a T-shaped piece of iron that is driven into the ground or into a board or block of wood and used as a back support for a piece of slate while the slate hammer chops the slate to size. Instead of a stake though, almost any straight edge, such as the edge of a ladder or even another slate, will do.

Few people realize that a hole can be punched easily in roof slate by using a sharp object such as a nail, and slates of standard 3/16" thickness do not need to be drilled. Traditionally, however, holes are punched in roof slate using a slate hammer, which has a sharp point on it for that purpose.

A crude way to *cut* slate is to use the pointed end of a slate hammer (or a hammer and nail) and poke holes along a line on the slate, then break the slate on the dotted line. Once the slate is cut, the jagged edge is tapped straight-on with the hammer to remove the roughness. In this manner, even hammers without cutting shanks can be used to cut slates. Also, this slate-cutting technique can be performed with *any* hammer and a nail, so that even the average do-it-your-selfer can cut slate without special tools.

MISCELLANEOUS HAND TOOLS

Other hand tools commonly required for work on slate roofs and frequently carried in a slater's tool belt are: **tin snips** (aviation snips) for cutting flashing and sheet metal; **utility knife** for general purposes; **tape measure** (25' length); **chalk line** (for chalking lines when installing new roofs or valleys); **utility pencil** (carpenter's pencil or other marker); **flat pry bar** (also called "wonderbar"), for removing exposed slate nails, prying up slates, removing ridge iron and flashings, scraping tar off roofs, etc.; and a **nail punch** such as a 1/2" bolt about 6" long, for setting the nail when replacing a slate (see Chapter 13), and for punching nails down through the slate.

An additional set of tools is required for masonry work on chimneys, and that is discussed in Chapter 15.

GROUND LADDERS

Not much work would be done on slate roofs without ground ladders. They're critically important pieces of equipment that have a degree of versatility to them, and, with experience, can be used to gain access to even the most unlikely of places. The recommended type of ladder is either a medium or heavy duty commercial, aluminum ladder. Many homeowners have household duty ladders, but these aren't recommended for slate roof work, because the ladder should be safe and sturdy, and household duty ladders tend to be flimsy. Speaking of safety, be sure to read the section pertaining to ladders in Chapter 10 (Safety First), and don't electrocute yourself when putting up or taking down an aluminum ladder.

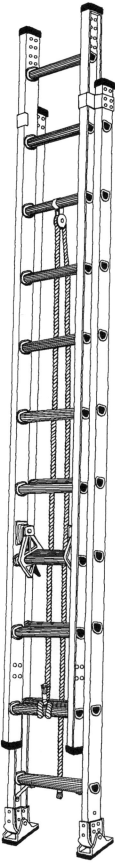

Figure 11.6 Ladders have a top and a bottom. The bottom has the feet (some people do put ladders up upside down!). The moveable part of the ladder is called the "fly."

The first task when using a ladder is putting it up. This can be done by one person by simply lodging the base of the ladder against a solid object such as the side of a house, and "walking" the ladder up (see Figure 11.8). To walk the ladder up, simply start at the top of the ladder, hoist it overhead at arms length, and start walking toward the stabilized base of the ladder, moving your hands forward rung to rung until the

Figure 11.7

Plumbing a Ground Ladder

Moving a foot backward or forward will drop or raise that foot slightly (depending on the make of the ladder), thereby helping to plumb the ladder.

ladder is standing straight up. Then pull the base of the ladder back away from the building so the ladder is standing vertical, extend it until it reaches the eaves, then lean it against the building. If you're walking up an extension ladder, do *not* extend the ladder *before* walking it up; instead, extend it after it has been stood up in a vertical position (this all may take some practice for the inexperienced person). The ladder is taken down in the same manner - the extension is retracted, the base is wedged against something solid and unmoveable, and the ladder is walked down backwards.

Once a ladder is standing vertically and not leaning on something, it's very unstable unless it's kept perfectly plumb (straight up and down). So it's extremely important to hold the ladder plumb when moving it around, such as alongside the house. Better to leave the top of the ladder leaning slightly against the building when moving it along, if possible.

It's easier to handle a ladder with two people. To put up a ladder with two people, one keeps a foot on the base of the ladder while the other walks it up. The ladder is turned on edge during this procedure to make it easier to keep a foot on the ladder (see Figure 11.10). The foot must not be taken off the ladder until the ladder is vertical, otherwise the base will fly up, the top of the ladder will plummet to the ground, and loud cursing noises will come from the guy trying to walk it up.

Once the ladder is up in a vertical position, one person stands behind the ladder (between the ladder and the building) and supports it with two hands, while the other extends the fly (moveable) section of the ladder, even climbing the ladder if necessary. The person supporting the ladder must be strong enough to hold it nearly plumb, but leaning a little toward the house, so it doesn't fall backward or sideways with the other person on it.

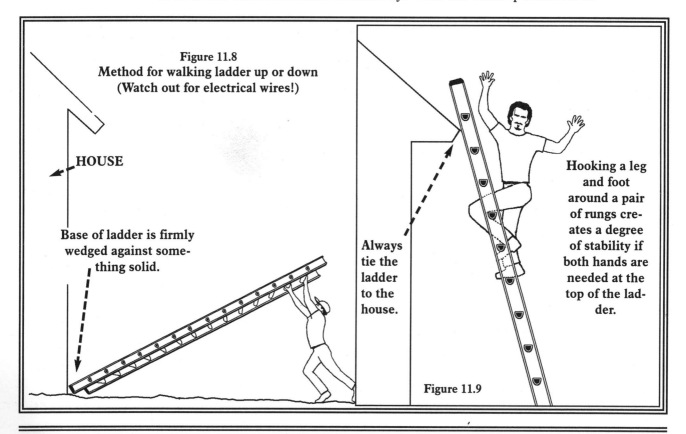

Figure 11.8
Method for walking ladder up or down
(Watch out for electrical wires!)

HOUSE

Base of ladder is firmly wedged against something solid.

Always tie the ladder to the house.

Hooking a leg and foot around a pair of rungs creates a degree of stability if both hands are needed at the top of the ladder.

Figure 11.9

Always watch for electrical wires when putting up or taking down a ladder!

Figure 11.10
It is much easier for two people to put up a ladder than one person. The fellow on the left keeps his foot planted <u>firmly</u> on the ladder until it is up <u>all the way</u> and he pulls on the ladder to help lift it as the other fellow walks it up. The ladder is turned on edge to allow for a secure place to set one's foot. The procedure in reverse will safely bring a ladder down.

If a person is putting a ladder up alone and must extend the fly, the ladder is held plumb with one hand, a foot solidly holds down the bottom rung of the ladder, and the other hand extends the fly either by pushing on it, or by pulling on the ladder rope, until the fly reaches the eaves of the building, if possible. Then the person can climb the ladder and extend the fly to the desired height by pushing the fly upward as the ladder is climbed, being careful that the ladder is leaning against the building with enough slope to keep it from falling backward. Once the ladder has been extended, the base can be moved back from the building to its proper final distance, which should leave a slope on the ladder that is not too steep or too shallow.

The ladder top should *always* be tied to the building after the ladder has been erected. Rain spouting often comes in handy as a place to tie on to, but when no rain spouting is available, something else must be used. A 16 penny nail driven solidly into the fascia and bent into a hook will make a quick anchor to tie on to when nothing else is available.

Many aluminum ladders come equipped with an attached rope meant to be used to extend the ladder. Some people swear by this rope, others take the rope off right away and never use it at all. Originally, extension ladders were made of wood and were quite heavy, and ropes weren't sufficient to extend them, so many roofers developed the habit of extending the ladder by climbing up the rungs and pushing the "fly" ahead of themselves. This requires two people when a ladder is long or must be extended very far, but can easily be done by one person in most situations.

Any time a ladder is stood up vertically and the base placed in its permanent location, the ladder will not stand plumb if the ground on which it is standing is not level. Before the ladder is leaned against the building it must be plumb in order for it to be safe, otherwise it may slide in one direction or another along the eaves or along the rain spouting, causing the ladder climber to lose balance. To check to see whether a ladder that is already leaning against a building is plumb, stand on the bottom rung and pull the top of the ladder away from the building slightly. A plumb ladder will drop right back against the building in the original place. The top of an incorrectly erected ladder will move to one side or another. It's always better to check a ladder for plumbness before you climb on it, especially if someone else put it up. You don't want to get to the top and realize the ladder hasn't been put up properly and the top is now sliding sideways, with you on it!

To make the ladder plumb on an uneven surface *don't* prop one of the legs on top of something if you can avoid it. Instead, *dig a hole* and drop the other leg into it. ***It is much more stable and safer to dig a hole for one ladder foot than to prop the other foot up***, and a hole can be dug quickly in soil with the claw end of a claw hammer, and easily restored after the ladder is moved.

This can't be done, of course, when on pavement, or setting a ladder up on a roof (porch roof for example), and in these cases a prop must be used to level the ladder's feet. A solid plank with one end propped on a block of some sort usually does the trick.

Most ladders have feet that swivel backward or forward. These come in very handy for leveling the base of the ladder when the surface is just slightly out of level. Many ladders will drop slightly on one side if the foot on that side is swiveled backward, and will raise slightly when the foot is swiveled forward. Some ladders do just the opposite, so you'll have to experiment with yours. It's routine to climb a ladder when its feet are bent (swiveled) backward or forward (see Figure 11.7).

The more extended the ladder is the more top-heavy it will be, and the less stable it will be when trying to move it. *The rule of thumb is to retract the fly before moving a ladder, then extend it again when the ladder's where you want it. The other rule of thumb when moving a ladder is to keep it as plumb as possible when moving it. Often it's best to walk the ladder down, move it, and walk it back up.*

Most of the work commonly done on slate roofs is done on hook ladders (roof ladders), not on ground ladders, unless working along the eaves (drip edge) of a roof. If it's necessary to work from the top of a ground ladder and both hands are required, leaving the worker on the ladder feeling unstable with nothing to hold on to, then one of the worker's legs should be wrapped around a pair of ladder rungs as in Figure 11.9.

A good all-around ladder for working on a two-story house is a 32 foot medium duty commercial ladder. A 40 foot ladder is necessary for higher homes, but a 32 foot ladder will easily reach well beyond the eaves of most homes, and it's a relatively easy ladder to handle, especially by one person.

HOOK LADDERS

A hook ladder is any ladder that has a ridge hook attached to the end of the ladder so the ladder can be hooked over the ridge of a roof. Hook ladders are one of the most important pieces of equipment for anyone wanting to work on a slate roof. They make a difficult and dangerous job relatively safe and easy, and they're not hard to come by because you don't have to buy a special ladder with a hook already attached to it. Instead you buy only the hook, separately, and attach it to one of your own ladders. Hook ladders are discussed in the previous chapter, and if you haven't read about them there already, now is a good time to do so.

Hook ladders enable the roofer to climb on slate roofs without putting weight on the slate itself, thereby preventing breaking the slate. They also keep the roofer off the slate when the roof is too hot to touch, thereby preventing the roofer from becoming roof-burned (it happens). *Never use ropes to climb on slate roofs* because they force you to walk on the slates, which are subject to breaking under the weight of a person. Slate roofs are not sidewalks and should not be walked on unless absolutely necessary. (One of the main reasons slate roofs go bad on porches and other low-slope roofs is because people tend to walk on them. Generally speaking, the steeper and higher a roof, the better condition it will stay in, since people stay off steep roofs.) In addition to ropes, home-made chicken ladders (wooden hook ladders) don't make any sense when steel ladder hooks are so inexpensive and safe.

A good ladder hook to use on slate roofs is the No. 606 Murray-Black ridge hook (see Figure 11.11). This hook has a flat piece of angle on the end which prevents the slate from breaking under the pressure of the hook. It's also easy to clamp onto a ladder, as it's simply locked into place by a wing nut. Although the hook is typically attached to the top two rungs of the hook ladder, it can be attached to any pair of rungs lower down on the ladder in situations

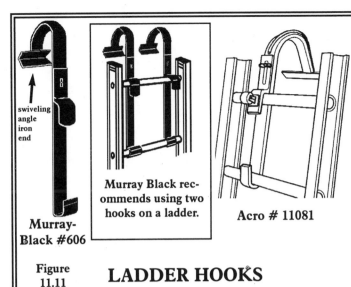

swiveling angle iron end

Murray Black recommends using two hooks on a ladder.

Acro # 11081

Murray-Black #606

Figure 11.11

LADDER HOOKS

Murray Black Co. (wholesale supplier), 1837 Columbus Ave., Springfield, OH 45503; (937) 323-3609; Ask for the #606 ladder hook. Check local roofing supply companies for retail sales of Murray-Black ridge hooks. This hook is inexpensive, well-made, and has an angle-iron end that swivels, making it a good hook for slate roof work.

Acro Building Systems, Inc. 2200 West Cornell, Milwaukee, WI 53209; Ph: (800) 267-3807 or (414) 445-8787; Their ladder hook is sturdy and has a two-way roller wheel making it easier to roll the hook ladder up the roof. The bar at the end of the hook doesn't swivel, however, making its usefulness on slate roofs limited. An improved design was planned as of 1997, and may be available.
(Also check local building supply outlets)

where the available hook ladder is too long to be practical (Figure 11.12).

There is another type of Murray-Black ridge hook: the 608 hook, which is essentially the same as the 606 hook, except that it has a rolling wheel which enables the hook to be slid up the roof more easily. *There is a major flaw in the design of this hook, however, in that the wheel swivels, allowing the top of the hook ladder to roll sideways on the roof. This is very dangerous when pushing a 20 foot long hook ladder up a steep roof - you don't want the ladder top to suddenly roll sideways when you just about have it hooked - it will go out of control and possibly knock you off your ground ladder.* This rolling hook would be greatly improved if the wheel only rolled in one direction (up and down the roof), and not sideways too! So if you have a hook with a swivel wheel on it (Murray-Black may have discontinued this hook by the time you read this), I'd suggest taking the wheel off *before* an accident happens. Don't confuse this dangerous hook with the Acro hook, which *does* roll properly.

It should be noted that the manufacturer of Murray-Black hooks specifies that two hooks be used at the same time, side by side, on the same rungs of each hook ladder (Figure 11.11). If you do any amount of slate work, however, you'll soon learn that a hook ladder with two Murray-Black hooks on it is very top heavy, hard to handle, and therefore a hazard of its own. The author

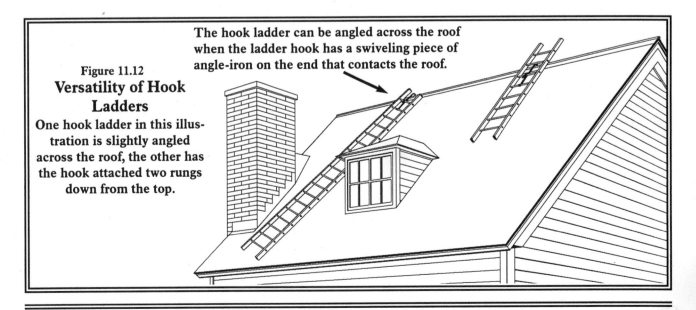

Figure 11.12
Versatility of Hook Ladders
One hook ladder in this illustration is slightly angled across the roof, the other has the hook attached two rungs down from the top.

The hook ladder can be angled across the roof when the ladder hook has a swiveling piece of angle-iron on the end that contacts the roof.

has never seen any roofer use more than one Murray-Black hook at a time on any hook ladder, but if you're going to follow the manufacturer's recommendations, you'll use two hooks.

Another ladder hook on the market is manufactured by Acro Building Systems, Inc. in Wisconsin (Figure 11.11). This is a sturdy hook, and unlike the Murray-Black hook only one hook is recommended per ladder. This hook comes with a rolling wheel that does not swivel, making it potentially a great ladder hook. Unfortunately, the bar at the end of the hook does not swivel either, and it should, which means it prevents the angling of a hook ladder across a roof, thereby severely limiting its usefulness. However, many tool manufacturers improve their tools as the demand calls for it, so by the time you read this, Acro may be marketing an improved hook. It might be worth looking into.

The disadvantage to a hook that can be fastened on and taken off the ladder easily is that the hook may come off when you don't want it to, like when you're sliding the ladder up the roof to hook it on the ridge. The author has *never* had this happen, but other contractors have had their hooks fall off. The solution to this potential problem is to *always* make sure the hook is attached square, snug, and centered on the rungs of the ladder and the wing nut is firmly secured. Obviously, if you put the hook on the ladder sloppily, you're just asking for trouble. The top edges of ladder rungs are slanted in one direction so as to allow for a level surface to stand on when the ladder is angled against a building. Hooks are designed to clamp onto those rungs and are made to accommodate that slant, so they *can* be put on backwards by someone who's not paying attention to what he's doing, and if so, *won't* attach firmly. This is one of the easiest mistakes to make when putting a hook on a ladder. When you're standing on a hook ladder on a high roof, the hook itself is the only thing protecting you from the force of gravity, a force to be reckoned with when working on steep slopes. That hook should be highly respected and used properly and with utmost care.

There are other styles of ladder hooks such as the one shown in Figure 10.4B in the last chapter (page 165), that are meant to be used in pairs because they attach to the side rails of the ladder and not to the rungs, and are used by people who only need to get up and down the roof (such as firemen) and don't need to actually work on a roof. But these are of no use to slate roofers.

In order to get a hook ladder up onto a roof, first the ground ladder must be set up properly. Then the hook ladder is "walked" up (as in Figure 11.8, often by propping against the base of the ground ladder) and laid against the ground ladder, leaving enough room to climb the ladder. The ground ladder is then climbed until the climber reaches about 3/4 of the way up beside the hook ladder, then he holds the hook ladder against his shoulder while hanging onto a rung tightly, and climbs up to the top of the ground ladder. At the top, the hook ladder is raised up far enough to tilt over the edge of the eaves and lay on the roof, afterwhich it is carefully (so as not to damage slate) slid up the roof, hook facing up, until the top of the hook ladder reaches the peak. The hook ladder is then flipped over and hooked on the ridge. The reverse procedure will get the hook ladder down. If the hook ladder needs to be moved sideways one way or the other, it can be "rolled" from the ground ladder (turned on its back then onto its front in a rolling motion).

It takes some practice to get hook ladders up and down safely, and it can be a dangerous job, especially on high, steep roofs requiring long hook ladders, or on windy days. Perhaps the most important trick in carrying hook ladders up to and down from roofs is to learn where the center of balance of the hook ladder is (hook ladders are top-heavy). Grab the ladder just above its center of balance when carrying it up or down a ground ladder. It will then tilt over the eaves and lay on (or come off) the roof easily. Make sure your body's chest level is just above the pivot point at the eaves of the roof when you tip the hook ladder on or off the roof. Once again, remember the rules of safety, and if it doesn't feel safe to you, don't do it - you can always practice with a shorter, lighter ladder first.

A professional slater may use several hook ladders at the same time on a roof. One may be used just to have a ladder to get up and down from the ground ladder to the roof peak, while another may be used to work the other side of the roof. Smaller hook ladders may be used to get around chimneys or to work on dormer roofs, and often hook ladders will be paired to make a roof scaffold as in Figure 10.5 (page 166). So it's always a good idea to have more than one hook handy, and more than one hook ladder available. Fortunately, any extension ladder can be taken apart and one half put to use as a hook ladder, then the extension ladder can be put back together when the hook ladder is no longer needed. This is one reason to remove ropes from ladders when not needed - then they can be taken apart and put back together more easily.

ROOF JACKS (BRACKETS)

Roof jacks are important tools for any roofer, especially slate roofers. Also called roof brackets, they're used to support planks on the roof and thereby create roof scaffolds, which allow a place to stand or to set another ladder on the roof surface. It's very important to know how to attach roof jacks to a slate roof safely without damaging the roof, as roof scaffolds will make many a seemingly impossible roof job vastly easier and safer. Speaking of safety, roof jacks are discussed in the previous chapter, and that information should be reviewed before attempting to use roof jacks. Roof jacks are also illustrated in Figures 10.6 and 10.7, page 167.

Many roof jacks fold up for storage, and this can be a hazard when the jacks have not been correctly latched open before being nailed on a roof. You don't want to get up on a roof scaffold and have it suddenly collapse because a roof jack wasn't locked open. Always check them

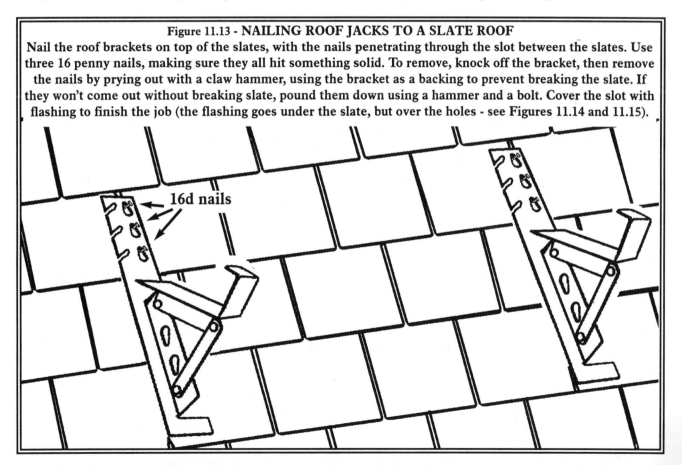

Figure 11.13 - NAILING ROOF JACKS TO A SLATE ROOF
Nail the roof brackets on top of the slates, with the nails penetrating through the slot between the slates. Use three 16 penny nails, making sure they all hit something solid. To remove, knock off the bracket, then remove the nails by prying out with a claw hammer, using the bracket as a backing to prevent breaking the slate. If they won't come out without breaking slate, pound them down using a hammer and a bolt. Cover the slot with flashing to finish the job (the flashing goes under the slate, but over the holes - see Figures 11.14 and 11.15).

16d nails

before putting your weight on them, or else use fixed (non-collapsible) jacks.

Roof jacks are typically nailed on *top* of the slate roof, in the *slots* between the slates where the slates abut one another side to side (see Figure 11.13), and are nailed *through* the slates. In the same way that a hammer and a nail can be used to punch a hole in a slate, a nail can usually be driven through a slate without cracking it. This is especially true of older slates, although some of the very hard slates, such as the New York red slates, may tend to crack when nailed through even after a century of age. For the most part, though, a nail can be driven through an old slate roof with impunity.

Roof jacks will always have at least three holes through which to nail them. Many have two sets of three holes. There is a good reason to have three holes on a roof jack - that's how many you should use when safely attaching the jack to the roof. One nail is not enough, two *may* do, but three is the right number. When nailing a roof jack to a roof always use three nail holes or slots and make sure all three nails hit something solid. If you're not hitting something solid with your nails, move the roof jack and try again (and be sure to cover your nail holes with flashing!).

Furthermore, on steep slope roofs use 16 penny common nails (3 1/2 inches long) when nailing roof jacks. When doing so, it's critically important to make sure the nail *heads* have properly hooked the roof jack, and are not driven in so far that the roof jack has nothing to hold on to (it's the nail *heads* that keep the jack from sliding off the roof, so use stout nails with good, large heads).

The jacks themselves only *hook* onto the nails, and they're readily removable with an upward tap of the hammer on the bottom of the jack, although the nails are another matter altogether. They must be removed from the roof or else covered with flashing after the work is done. Sometimes 16 penny nails can be pried from the roof with a claw hammer using a slate ripper or the jack itself as a backing to prevent the pressure of the hammer heads from breaking the slate. If this won't work, then the nails are simply driven into the roof using a nail punch (a 1/2" x 6" bolt will do fine), and left there.

After removing the roof jack, you'll have three nail heads or nail holes (depending on whether you pulled the nails out or drove them in) situated in a slot on the roof, and they'll surely leak. This is not a problem, however, as a piece of metal flashing called a *bib* is simply slid *under* the slot and *over* the holes to leave the area leak-proof. The metal should be non-corrosive (copper is ideal, but aluminum, especially *brown* painted aluminum, works well too), and should be bent lengthwise slightly so as to be force-fit under the slate, thereby preventing it from sliding back out. A lengthwise bend in the middle of the metal (see Figure 11.15) not only wedges the metal in place, but helps the metal ride over the nail heads, if there are any. The insertion of this metal flashing is almost always assisted by a slate ripper, which pushes the flashing in place.

The metal flashing should be about 4" wide and long enough to cover the nail holes with at least an inch of overlap on the bottom, and 2" of underlap under the next course of slates above (Figure 11.14). A length of 7" works well on many roofs, although sometimes longer flashing is required. *Don't use shiny flashing, as it looks unsightly in the cracks between the slates.*

Using the above technique, roof jacks can be safely attached to and removed from slate roofs without creating leaks.

SCAFFOLDS

The roof scaffold is only as good as its weakest part. Even if the roof jacks are nailed in place with three 16 penny nails each, and (if collapsible) firmly latched open, a bad plank will

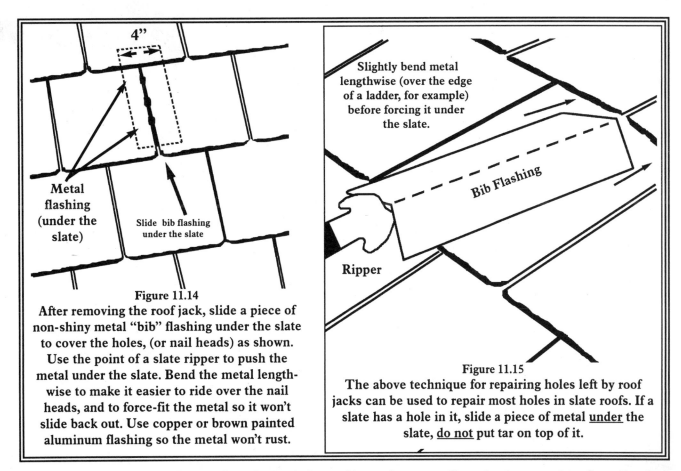

Figure 11.14
After removing the roof jack, slide a piece of non-shiny metal "bib" flashing under the slate to cover the holes, (or nail heads) as shown. Use the point of a slate ripper to push the metal under the slate. Bend the metal lengthwise to make it easier to ride over the nail heads, and to force-fit the metal so it won't slide back out. Use copper or brown painted aluminum flashing so the metal won't rust.

4"

Metal flashing (under the slate)

Slide bib flashing under the slate

Slightly bend metal lengthwise (over the edge of a ladder, for example) before forcing it under the slate.

Bib Flashing

Ripper

Figure 11.15
The above technique for repairing holes left by roof jacks can be used to repair most holes in slate roofs. If a slate has a hole in it, slide a piece of metal <u>under</u> the slate, <u>do not</u> put tar on top of it.

make the scaffold unsafe. A 2"x10" plank free of large knots, splits, checks, or other flaws in the wood makes a good scaffold platform piece. Typically, two roof jacks are sufficient for a plank that's eight feet long, although three roof jacks over a span of eight feet are much safer when a lot of weight is involved. If you doubt the strength of your roof scaffold, you can always add another roof jack to beef it up.

Other types of scaffolding may also be useful when doing work on slate roofs, and these include ground scaffolding (pipe or frame scaffolding) which sits on the ground and stacks one stage on top of another; and pump jacks, which attach to vertical 4x4's. Ground scaffoldings usually won't be required when doing general slate work, although they can come in real handy at times. Frame scaffolding especially can greatly enhance the safety of certain (usually high) jobs. Some roofers use ground scaffolding routinely when working on the drip edges and eaves of slate roofs, or when installing new roofs. Frame scaffolding can readily be rented.

SLATE HOOKS

Slate hooks are widely used by slate roofers to attach replacement slates in place, instead of using the nail and flashing technique discussed in Chapter 13, and no list of slate equipment would be complete without the slate hook. The slate hook is a simple copper or stainless steel hook that is nailed into the roof after a bad slate has been removed. The new slate is slid into place and the hook keeps it from sliding out of the roof. The copper hooks can be difficult to use on roofs that have hardwood sheathing because they may bend when you try to nail them in place, but some slaters swear by them and they are used around the world (Figure 11.16).

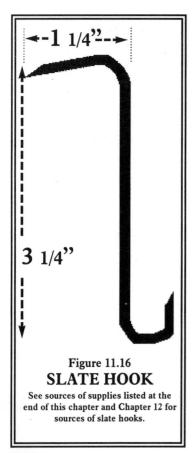

←–1 1/4"–→

3 1/4"

Figure 11.16
SLATE HOOK
See sources of supplies listed at the
end of this chapter and Chapter 12 for
sources of slate hooks.

The standard nails for nailing standard thickness (3/16") slate are one and a quarter to one and a half inch long *hot-dipped* galvanized (zinc coated), *or* copper roofing nails. These nails will last 100 years easily, and may last much longer. Most century old nails of these types are in good enough condition to be used over again. Longer copper nails are required for thicker slates.

Do not use *electroplated* galvanized nails, which are cheap nails made for asphalt shingles, and should never be used on a slate roof. Electroplated nails are marked "EG" roofing nails (electro-galvanized), whereas hot-dipped nails are clearly marked "hot-dipped." Copper roofing nails are the preferred nails specified by professionals in the slate and in the roofing supply industries. However, 99% of old slate roofs do not have copper nails, they have hot-dipped galvanized nails, which are far less expensive than copper and work almost as well. So even though industry professionals insist on using copper nails, if you're using a good, hot-dipped galvanized nail to fasten the slates to a roof, you'll find that they work quite admirably. Copper nails, however, *are* superior to hot-dipped galvanized; and stainless steel or aluminum nails may also be used.

The traditional nail for fastening ridge iron to a slate roof is the eight penny (2 1/2" long) hot-dipped galvanized nail. After this nail is pounded into place, the head is covered with a dab of roof cement or caulk (50-year durability clear silicon is recommended). Some roofers insist on using gasketed nails on ridge iron, but a standard eight penny nail with a caulked head works best.

Finally, when nailing copper (flashings or valleys) one should always use copper or brass nails to prevent galvanic action. Galvanic action occurs when dissimilar metals are placed in contact with each other, leading to the deterioration of the electropositive metal. This is discussed further on page 242.

ROOF CEMENT AND CAULK

Roof cement, also known as "tar," is a black, plastic material that comes trowel grade (thick) or brush grade (thin). For our purposes, the trowel grade material is much more useful. It also comes in two varieties - standard or "wet surface." Wet surface roof cement can be used to seal leaks on damp surfaces, such as during a rainstorm, and it's a good idea to keep some handy, although standard roof cement has more body and is the preferred material for general roof repairs. Roof cement is also available in caulking tubes.

A good all-around caulk for slate roofs is clear, 50-year silicon caulk. It really sticks to just about anything, holds tightly, lasts a long time, and is waterproof immediately (although it can not be used on damp surfaces). It's used to seal exposed nail heads, and to fill mortar joints on chimneys after they're reflashed.

LADDER JACKS

Ladder jacks (also called ladder brackets) are shown in Figure 10.5 , page 166. Like roof jacks, their purpose is to support planks and create working platforms off the ground. There are many different types of ladder jacks, and they're typically used most often by contractors who attach them to ground ladders for the purpose of working on the siding, gutters, soffit, fascia, or eaves of a house, but since this is a book about slate roofs, we will only discuss their practical value for slate roof work.

Ladder jacks are most useful to slaters when attached to hook ladders to create a quick and sturdy roof scaffold without nailing anything into the roof. As such, the best ladder jacks to use are lightweight aluminum ladder jacks, which can be carried up and down a ladder with ease. They require no tools to attach to the ladder, and simply clamp onto the ladder by design. They must be used in pairs, and therefore require pairs of hook ladders. They are especially handy when working on chimneys. Hook ladders must be heavy duty if the roof scaffold attached to them is going to support heavy weight, such as bricks.

Sources of Tools and Equipment (1997)

•**SLATE HAMMERS, RIPPERS, CUTTERS, STAKES AND MISC.:**
Techni-Tool (CK hammers), 5 Apollo Road, Plymouth Meeting, PA 19462-0368; Ph: (610) 941-2400.
John Stortz and Son, 210 Vine Street, Philadelphia, PA 19106; Ph: (215) 627-3855, Source of slate hammers, rippers, stakes, and cutters. Stortz tools (including cutters) can be bought at various retail supply outlets, including:
Evergreen Slate Co., 68 Potter Ave., Granville, NY 12832; (518) 642-2530 (They also have Esco cutters).
Camara and Sons Slate Quarriers, Hampton, NY 12837; (518) 282-9646.
Hilltop Slate Co., Route 22-A, Middle Granville, New York, 12849; (518) 642-2270.
New England Slate Co, Burr Pond Road, Sudbury, VT 05733; (802) 247-8809; They also have the New England Slate Bag for carrying slate.
Also check roofing supply companies, such as:
McClure-Johnston Co., 201 Corey Ave., Braddock, PA 15104; Ph: (800) 232-0018 or (412) 351-4300.
Cassady-Pierce Co., 2295 Preble Ave., Pittsburgh, PA 15233; Ph: (412) 321-8987, (800) 227-7239.
•**SLATE CUTTERS:**
Stortz cutters: John Stortz and Son (as above).
Esco slate cutter: Evergreen Slate Company (as above).
Gundlach slate cutter: Beno J. Gundlach Co., 211 North 21st St., Belleville, IL 62222; (618) 233-1781.
Ed Hack Cutter: This is a cutter similar to the Gundlach cutter (no hole punch), and may still be available through New England Slate, Burr Pond Road, Sudbury, VT 05733; (802) 247-8809.
•**RIDGE HOOKS:**
Murray-Black Co., (wholesale supplier) 1837 Columbus Ave., Springfield, OH 45503; (937) 323-3609; Ask for the #606 ladder hook. Check local roofing supply companies for retail sales of Murray-Black ridge hooks.
Acro Building Systems, 2200 W. Cornell Street, Milwaukee, WI 53209; Ph: (800) 267-3807 or (414) 445-8787.
•**ROOF JACKS, LADDER JACKS, LADDERS, AND MISC. TOOLS:** (These tools can be bought at almost any local building supply store, hardware store, or roofing supply outlet.)
•**ALUMINUM LADDER JACKS:**
Murray-Black, 1837 Columbus Ave., Springfield, OH 45501; Model A-615; Ph: (937) 323-3609.
Acro Building Systems (as above), Model #11753 (3 rung) or #11752 (2 rung).
Werner Co., 93 Werner Rd., Greenville, PA 16125-9499; (888) 523-3372 or (412) 588-8600; They have a whole line of aluminum ladders and accessories including three types of aluminum ladder jacks: 2 rung short body #10-14-2, 2 rung long #10-20-2, 3 rung long #10-20-3.
•**SLATE BRACKET:**
Acro Building Systems (as above), Model 19600 is a 6" wide roof bracket useful for stacking slate on a roof.
• **LADDERS:**
Werner Co., 93 Werner Rd., Greenville, PA 16125-9499; (888) 523-3372 or (412) 588-8600; They have a whole line of aluminum ladders and accessories.

Note: If you're a tool or equipment manufacturer who makes tools suitable for slate roof work and are not listed above, please contact the author c/o the publisher listed at the front of this book.

Chapter Twelve

Installing and Recycling Roof Slate

Roof slates in the United States are traditionally nailed onto solid wood roof decks built of the lumber derived from local forests. This traditional style of construction has been tried and proven over centuries, and should be duplicated when installing slate roofs on new or restored construction today. The primary characteristics of traditional American slate roof construction are as follows: 1) The slates are nailed onto a completely sheathed roof deck; 2) Full one inch thick, local, green or air-dried lumber boards are used for roof sheathing (optionally covered with 30 lb. felt paper); and 3) The slates are nailed onto the wooden sheathing using two nails per standard slate (no adhesives are used).

1) Solid Wood Roof Sheathing

The slates are nailed onto a completely sheathed roof deck - the roof deck is constructed of full boards abutting each other so as to cover the entire roof surface, leaving no, or negligible, gaps between the boards..

This is in contrast to the practice of nailing the slates onto "slater's lath," which are narrow strips of wood spaced evenly on top of, and perpendicular to the roof rafters so as to provide a place to nail the slate (Figure 12.1). Roof lath construction only requires the bare minimum of wood sheathing - just enough to provide an anchor for the nails, and is therefore more conservative in the need for lumber. This is the preferred method of construction in Wales, where virtually all of the traditional buildings are constructed of stone or slate walls, and wood is minimally used. American barns also tend to have lath-type roofs, no doubt because of the lesser cost of the

Figure 12.1 Slating lath roof construction on old farm building, Yorkshire-Lancashire Border, England. Closeup on next page.
(Photo by Dave Starkie)

materials.

The practice of using roof lath to support slate may have developed as a response to the scarcity of forest resources in Wales and Europe in the 18th and 19th centuries. That scarcity did not exist in the United States during that time, when someone in America came up with the idea that the roof deck should be completely covered with wood, rather than with just strips of lath. This was a good idea - it created a stronger, better insulated roof that could withstand the snow weights typical of the American northeast. A stronger roof makes for a stronger building, and America's wooden building styles benefited from a roof of solid sheathing. Solid sheathing also provides much more nailing surface, making it easier to repair slate roofs, and therefore facilitating their restorability.

It's interesting to note that roof slates in Wales were originally "hung" on roof lath using wooden pegs driven through the slates. The lath were originally stripped, not sawn, from local oak and pine logs. The pine pegs were split down to the size of a pencil stub from a block of wood, and a single peg was driven through the top center of the slate so that the peg was flush with the front of the slate, which was then hung on the lath (Figure 12.3). The heavy weight of the slates overlapping each other held the roof together.

Today, the Welsh use a modified version of the lath and peg system, which we will look at later in this chapter.

2) Use Local Lumber

Full one inch thick, local lumber boards are used for roof sheathing, unplaned, not tongue-in-groove, and not kiln dried. This is in contrast to today's architectural recommendations to use laminated wood (such as plywood) under slate, which is

Figure 12.2
▲ **Close-up of lath and peg roof construction on old farm building in England.** (Photo by Dave Starkie)
▼ **Solid sheathing made from local mixed hardwoods only one week "off the stump." This PA shed roof will be slated with recycled sea green slates.**
(Photo by Bob Sayre)

not recommended by the author.

Construction methods in America follow fads, such as the ventilated ridge fad now popular among builders, discussed in the next chapter. One of the most persistent fads involves the use of laminated woods and kiln-dried wood imported from great distances to the northeastern United States for construction purposes. This is ironic because the northeast is a great forested area abundant in tree species that yield very high quality wood, wood that is excellent for such things as roof decks. Even more ironic is the fact that the superior woods of the American northeast are less expensive than the imported kiln-dried and laminated lumbers that come from the northwest and from Canada. When you add to these facts one more, that virtually all of the older homes in the northeast were built from local lumber, which has been tried and proven over the centuries, one must wonder at the adamant objections some modern carpenters have about using local lumber to build roofs.

Do builders object to using local lumber? Yes. In fact, some will adamantly insist that "It can't be done!" Contractors often pursue the easiest and most convenient methods of construction in order to maximize their "bottom line." Their goal is to get the construction job done and get paid, and they know full well that someone else will have to worry about repairing the roof when the plywood delaminates years later -- it won't be their problem. They therefore have a vested interest in convincing home-owners to go with slap-it-down plywood on their roofs, rather than the proven traditional slate roof sheathing made from full boards, usually obtained locally. And since most home-owners don't have a clue about building materials and methods, they're easy prey to smooth talking contractors. Homeowners want to trust their contractors, but if you're a homeowner and your contractor is insisting you must use plywood under your slate, he's wrong.

Figure 12.3

▲ A slate from a 16th century abbey in Wales showing the actual peg and lath used in the construction of the roof.

▼ A hundred + year old American barn roof with white oak lath construction. Note the deplorable condition of the disintegrated terne metal valley, 34 feet long. The oak lath was still solid, however, making the valley replacement a routine job. Imagine the nightmare of a repair job if the barn had been sheathed in plywood, which would have delaminated the entire length of the valley!

(Top photo by the author, bottom photo by Umberto Perlino)

Modern contractors have accepted the notion that roofing is temporary, and many of these builders can't comprehend the idea of a permanent roof. Slate roofs, however, should always be built with a 100 year lifespan in mind (for soft slate as well as recycled hard slate) and a 200 year lifespan in mind for new, hard slate. The entire concept of a roof lasting a century or two is foreign to modern American builders, and they can be reluctant to deviate from their normal construction techniques (i.e. particle-board or plywood roof decking). They may even talk a home-owner out of a new slate roof based on the contention that they can't use their preferred low-quality roof deck material (I've seen them do it!).

Green lumber is not kiln dried, and may not even be air dried, but it still makes a great roof deck. The boards simply dry in place after being nailed to the rafters. When fully dry, they'll have a gap between them of about 1/2 inch on the sides (wood shrinks more width-wise than lengthwise by a factor of seven). Contractors will insist that the boards will warp, crack, and twist when they dry, but these are contentions based upon inexperience. When nailed properly, the boards remain flat.

In fact, many hardwoods such as oak *must* be used green. When completely dry they're too hard -- nails won't penetrate them without bending. It's worth mentioning that a tree felled in the winter when the sap is down will dry more quickly and shrink less than one felled when the sap is up.

A one inch thick, solid oak roof deck is a strong, durable one which will last for centuries. And if a leak damages part of the roof sheathing, it can simply be cut out and replaced with the same material - local, green lumber. Plywood, on the other hand, does not have the test of time behind it, and far too often ply-

Table 12.1
GENERAL CHARACTERISTICS OF WOOD

1) Degree of workability with hand tools; 2) Tendency to warp; 3) Tendency to shrink or swell; 4) Relative hardness; 5) Comparative weight

H = High, I = Intermediate, L = Low

Species	1	2	3	4	5
Black Ash	L	I	H	H	I
White Ash	L	I	I	H	H
Basswood	H	I	H	L	L
Beech	L	H	H	H	H
Yellow Birch	L	I	H	H	H
Eastern Red Cedar	I	L	L	H	H
Western Red Cedar	H	L	L	L	L
Northern White Cedar	H	L	L	L	L
Southern White Cedar	H	L	L	L	L
Cherry	L	L	L	H	I
Chestnut	I	L	L	I	I
Cottonwood	I	H	H	L	I
Southern Cypress	I	I	I	I	I
Rock Elm	L	I	H	H	H
Soft Elm	L	H	H	H	H
Balsam Fir	I	I	I	L	L
Douglas Fir	L	I	I	I	I
White Fir	I	I	I	L	L
Red Gum	I	H	H	I	I
Eastern Hemlock	I	I	I	I	I
Western Hemlock	I	I	I	I	I
Pecan Hickory	L	I	H	H	H
True Hickory	L	I	H	H	H
Western Larch	L	I	I	H	H
Black Locust	L	L	I	H	H
Honey Locust	L	L	I	H	H
Mahogany	I	L	L	H	H
Hard Maple	L	I	H	H	H
Soft Maple	L	I	I	H	H
Red Oak	L	I	H	H	H
White Oak	L	I	H	H	H
Ponderosa Pine	H	L	I	L	I
Arkansas Soft Pine	H	L	L	L	L
Sugar Pine	H	L	L	L	L
N. White Pine	H	L	L	L	L
W. White Pine	H	L	L	L	I
S. Yellow Pine	L	I	I	H	H
Yellow Poplar	H	L	L	L	L
Redwood	I	L	L	I	I
E. Spruce	I	L	I	L	I
Sitka Spruce	I	L	I	L	I
Sycamore	L	H	H	H	H
Tupelo	L	H	I	H	H
Walnut	I	L	I	H	H

(Source: H. E. Brosius Co., Kittanning, PA; Home Handbook (1956), p. 4)

Table 12.2 LUMBER GRADES Common or Board Lumber	AVERAGE WEIGHT OF TIMBER		APPROX. WEIGHT OF DRY LUMBER PER 1000 BOARD FEET	
	Species	**Weight (lbs/cu. ft)**	**Species**	**Weight (lbs)**
No. 1: Best quality, most expensive. generally clear of knots, but may have small, tight knots.	Ash	42	Ash 3,500	
	Chestnut	41	Chestnut 3,400	
	Hemlock	25	Hemlock 2,100	
No. 2: All around utility grade. Ideal for roof construction. May have occasional wood defects that need to be cut out.	Hickory	53	Hickory 4,400	
	Maple	49	Maple 4,100	
	Oak	32-48	Oak 4,000	
	Norway Pine	36	Norway Pine 3,000	
No. 3: May have knot holes. Lower grade than #2.	N. Yellow Pine	34	White Pine 2,100	
	S. Yellow Pine	45	Yellow Pine 3,000	
No. 4: Lowest grade generally considered usable for construction of any kind.	Spruce	25	Spruce 2,100	
	Walnut	48	Walnut 4,000	
No. 5: Not suitable for construction.	(Source: Radford (1913), p. 208)		(Source: Radford (1913), p. 209)	

wood is found de-laminating on older roofs. The author once tore a 20 year old plywood roof apart on a suburban home, and found the bottom three feet of the plywood to be almost completely delaminated. Today it's recommended that all plywood be covered with a 3' wide weatherguard plastic contact paper along the eaves of a building to prevent delamination; so now contractors use lumber filled with glue and then coat it with plastic to make it work. Why bother? Use a solid board, especially from a local source - it works in its natural state, it's less expensive, purchasing it feeds the local economy, it's ecologically preferable to imported and laminated wood products, it gets the builders in touch with their local "bioregion" by using local tree species, and it follows tradition in the construction of slate roofs.

Local lumber can be air dried, which simply means that the lumber is "stickered" for a few weeks or months before using. Stickering lumber means stacking it with dry sticks of uniform thickness (about an inch) between the layers of boards so air can circulate around the lumber, thereby drying it out. However, when building, framing, or sheathing with green lumber, stickering is not necessary if the lumber will be nailed in place within two weeks of cutting. If the lumber is going to sit around longer than that, it should be stickered and protected from rain until it's time to use it, otherwise it may develop surface mold.

As with slate, building with local lumber is an art from a by-gone day. The author has had the pleasure of building with local lumber for twenty years at the time of this writing, and virtually every slate roof the author has worked on sat atop a house built of local lumber. When these roofs are repaired or restored, local lumber is used as needed. When new slate roofs are built, they're built from local lumber.

Local lumber is usually only sold by the sawyers who saw it; it usually needs to be ordered in advance and the buyer may wait for weeks when large quantities of specific species are ordered. It's also "green" (undried) when you buy it. Furthermore, it's heavier than the kiln-dried softwoods typically sold

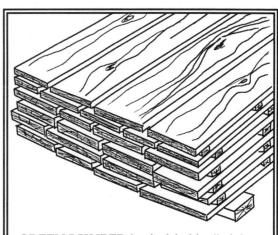

GREEN LUMBER is air dried by "stickering" it - stacking it with dry, one inch wooden spacers between rows of boards, then covering it to keep the rain off.

by building material centers, and therefore it requires more effort to work with. Because of the extra effort in locating a sawyer, ordering the lumber, waiting for it, hauling it, stickering it (if necessary), then lugging it up on the roof, only the most conscientious of builders use it nowadays. It is well worth finding those builders when installing a slate roof if you want the roof to last. If a contractor tells you he can't use local lumber because it's green and it will twist and warp, he's throwing you a curveball. Find another contractor. And it's well worth locating a local sawyer and ordering your roof decking directly from him if you're going to build your own roof.

You can locate local sawyers by looking under "lumber-wholesale" in the yellow pages of your phone book. Rough sawn lumber is full size - a 1x8 is a full one inch by eight inches, therefore, the "board feet" required for the job is actually the square feet of roof area including overhang. When ordering local lumber, always allow for *at least* 15% waste when figuring your total area, and more when valleys and hips are being built. Don't deduct from the total area for skylights or chimneys.

Use #2 grade, 1" thick boards at least 6" wide. Preferred species are the softwoods such as pine and hemlock, or the hardwood yellow poplar (tulip poplar) because they're lighter and therefore easier to work with, although almost any tree species suitable for lumber will do, including oak, cherry, maple, ash, birch, and beech (see Tables 12.1 and 12.2). The boards do not need to be planed or joined, or milled in any way, although *it is important to get boards from a sawyer who does a good job of cutting his lumber*. Some sawyers are sloppy and their boards will vary as much as 1/2" (or more) in thickness or width from one end to another. If you do run across a bad board in your lumber, set it aside and use it for something else - not for the roof. If you find a board with an excessively large knot or other defect, cut the defect out before using the board, or don't use it at all. This is one of the simple tricks to using local, green lumber - sorting out the bad boards, and you probably will have some. It's the boards with the bad knots that are likely to warp when drying. When nailing the boards to the rafters, butt them firmly against each other, both at their sides and at their ends, and nail them down tightly. The slates are nailed to the roof sheathing while green; you should not wait for the wood to dry.

A suitable alternative to using local lumber for the roof decking under slate is to use 3/4" thick, planed, kiln-dried roof sheathing boards available from any lumber yard. These are usually made of pine and can be quite a bit more expensive than local green lumber, but are certainly preferable to plywood.

Old roofing books call for tongue-in-groove roof decking under slate. Yet, after a hundred years, tongue-in-groove boards shrink enough that the tongue and grooves are no longer joined. Slate roofs get very hot under the sun, and lumber will completely dry out under slate. Even kiln dried lumber will significantly shrink on a slate roof. The best lumber for your money will be the traditional, local, green, rough-sawn roof sheathing. One of the longest lasting roof sheathing lumbers is white oak, although it is heavy and somewhat difficult to work with, and hard to nail into when dry.

Nail the roof sheathing to the rafters with eight penny common nails (2 1/2 inches long). Cover the sheathing, no matter how green, with one layer of 30 lb. roofing felt paper, overlapped about three inches at the top edges, and nailed to the roof with 1" hot-dipped galvanized roofing nails. (The Welsh, incidentally, install their roofing paper *under* the slating lath.) If the felt paper must act as a temporary roof for a prolonged period (for example, over winter), then skim over each and every nail head with a little roof cement. This is not as big a job as it sounds, and it will keep the water out of the roof until the slate is installed.

In fact, felt paper isn't absolutely necessary for the roof to function; many slate roofs (primarily barn roofs) don't have any felt paper at all. Felt paper does, however, provide a temporary cover in the event of rain during installation, and it helps insulate and waterproof the roof, so it is recommended to use 30 lb. felt in order to do the best job.

3) Nailing the Slates

The roof slates are nailed onto the wooden sheathing using two nails per standard slate, the nails being approximately 1 1/4" - 1 1/2" long and typically made of *hot-dipped* galvanized (zinc-plated) steel. Alternatively, hardened copper nails of the same length are used, which are the preferred nail on the more expensive homes and buildings with slate roofs. However, most existing slate roofs are fastened with galvanized nails, and providing that they are hot-dipped galvanized, they seem to hold up quite well in most cases. Stainless steel and/or aluminum nails may also be used. Roofs with thicker slates require longer nails.

American roof slates come from the quarry already pre-punched for nail holes, in contrast to Welsh roof slates, which come from the quarry without holes. American roofs tend to have solid sheathing, whereas Welsh roofs tend to have slating lath, so American slates can be pre-punched without worry as to whether the nail holes will line up with the lath. Welsh roofers must "hole" their slates on the job site.

Nail holes should just clear the top of the underlying slate (Figure 12.4 below). This allows the slate to lay best. If the holes are too low you'll nail through the top of the underlying slate, which should not be done. Why not? Because it's much more difficult to remove and replace "double-nailed" slates due to the inaccessibility of the upper pair of nails. On the other hand, if the regular nails are nailed too far up on the slate, the slates may not lay flat on the roof.

The slates should not be nailed down tight, but are actually "hung" on the roof nails, so they can "float" on the roof. This prevents the slates from being subjected to damaging pressures which can crack the slate over time. It takes some practice to get the hang of nailing roof slates properly (no pun intended), but it isn't difficult. The nail holes are pre-punched so that the beveled side of the hole faces the front of the slate, allowing for the nail head to set down into the beveled hole, thereby preventing the nail from rubbing against the slate over top of it, and wearing a hole in it. When poking a hole in a slate, the strike against the slate should be against the back, so the bevel appears on the front (also see Figure 12.5 below, and Figure 5.11, page 88).

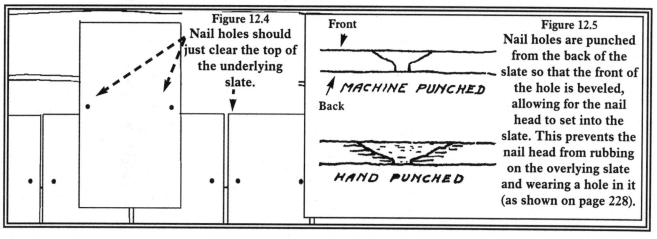

Figure 12.4
Nail holes should just clear the top of the underlying slate.

Front
MACHINE PUNCHED
Back

HAND PUNCHED

Figure 12.5
Nail holes are punched from the back of the slate so that the front of the hole is beveled, allowing for the nail head to set into the slate. This prevents the nail head from rubbing on the overlying slate and wearing a hole in it (as shown on page 228).

Table 12.3
APPROXIMATE WEIGHT OF SLATE PER SQUARE FOOT

Thickness (inches)	1/8	3/16	1/4	3/8	1/2	5/8	3/4	1
Weight (pounds)	1.80	2.70	3.62	5.47	7.25	9.06	10.9	14.5

[Source: Radford's Estimating and Contracting, 1913, p. 253]

AVG. WEIGHT OF NEW SLATE
100 Square Feet

Thickness	Weight (lbs)
3/16"	700-750
1/4"	1,000
3/8"	1,500
1/2"	2,000
3/4"	3,000
1"	4,000
1 1/4"	5,000
1 1/2"	6,000
1 3/4"	7,000
2"	8,000

[Source: Slate Roofs (author unknown), 1926, page 13]

WEIGHT OF SLATE PER 100 SQUARE FEET

Slate Length (in.)	Weight in Pounds/Square for Thickness (in.) Shown							
	1/8	3/16	1/4	3/8	1/2	5/8	3/4	1
12	480	725	968	1450	1938	2420	2900	3870
14	460	688	920	1370	1845	2300	2760	3685
16	445	668	890	1336	1785	2230	2670	3568
18	435	650	870	1305	1740	2175	2608	3480
20	425	638	850	1276	1705	2130	2555	3408
22	418	625	836	1255	1675	2094	2508	3350
24	412	616	825	1238	1655	2066	2478	3307
26	408	610	815	1222	1630	2039	2445	3265

[Source: Radford, Wm. A. (1913), Radford's Estimating and Contracting; Radford Architectural Co., Chicago, p.253]

RECOMMENDED RAFTER SIZES FOR RIGID CONSTRUCTION

Rafter spacing, (inches on center) ➤	12"	16"	24"

Unsupported Span ▼

[when 50 lbs/ft^2 live load (e.g. snow weight) anticipated]

	12"	16"	24"
6 ft.	2x6	2x6	2x6
8 ft.	2x6	2x6, 3x6	2x8, 3x6
10 ft.	2x8, 3x6	2x8, 3x6	3x8, 2x10
12 ft.	2x8, 3x8	2x10, 3x8	2x10, 3x10
14 ft.	2x10, 3x8	2x10, 2x12	3x10, 3x12
16 ft.	2x10, 3x10	3x10, 3x12	3x12, 2x14
18 ft.	3x10, 3x12	2x12, 3x12	2x14, 3x14
20 ft.	2x14, 3x12	3x12, 3x14	3x14
22 ft.	2x14, 3x14	3x14	
24 ft.	3x14	3x14	

[when 30 lbs/ft^2 live load anticipated]

	12"	16"	24"
6 ft.	2x4	2x4,2x6	2x6
8 ft.	2x6	2x6	3x6, 2x6
10 ft.	2x6, 3x6	2x8, 3x6	2x8, 3x6
12 ft.	2x8, 3x6	3x8, 3x8	2x10, 3x8
14 ft.	2x8, 3x8	2x10, 3x8	2x10, 3x10
16 ft.	2x10, 3x10	2x10, 3x10	2x12, 3x12
18 ft.	2x10, 3x10	2x12, 3x12	3x12, 2x14
20 ft.	2x12, 3x12	2x12, 3x12	3x12, 3x14
22 ft.	2x14, 3x12	3x12, 3x14	3x14
24 ft.	3x12, 3x14	3x14	3x14

Rafter sizes can vary depending on strength (species) of wood.
[Source: Slate Roofs (1926), author unknown, pp. 36-39.]

Figure 12. 6
RECOMMENDED ROOF SLOPES FOR SLATE ROOFS
(WITH MINIMUM HEADLAPS)

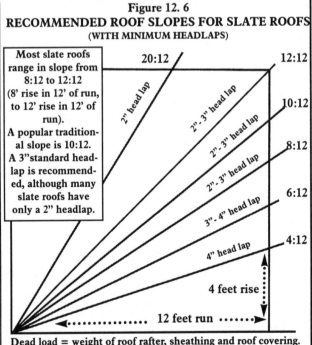

Most slate roofs range in slope from 8:12 to 12:12 (8' rise in 12' of run, to 12' rise in 12' of run). A popular traditional slope is 10:12. A 3"standard head-lap is recommended, although many slate roofs have only a 2" headlap.

Dead load = weight of roof rafter, sheathing and roof covering. Live load = additional total weight roof may be subjected to, such as from snow or wind. According to Slate Roofs (1926), roofs having a rise of 4 inches or less per foot of run, shall be assumed to have a vertical live load of 30 lbs. per square foot. A slope of more than 4:12, but less than 12:12 shall be assumed to have a live load of 20 lbs. per square foot. Slopes exceeding 12:12 shall be assumed as having no vertical live load, although provisions must be made to compensate for a wind force of 20 lbs per square foot. In localities where snow loads are an important consideration, the loadings shall be increased in accordance with local experience.

FRAMING THE ROOF

Slate is heavy - standard 3/16" thick slates weighs about 700 pounds per square (100 square feet of finished roof) when new. This is about three times as much weight as standard asphalt shingle roof material, and it raises concerns among people who are afraid their structure won't stand up to the heavy load of slate. How should a slate roof be framed? What are safe timber sizes and spans?

Many older homes have 2x10 rafters on two foot centers (usually hemlock, a local softwood), and this has proven to work well. However, rarely is a 16' span on a rafter left unsupported without an inside wall or collar brace, and it's recommended that a 2x10 rafter on two foot centers not exceed 10' in free (unsupported) span (see Figure 12.6). Many older slate roofs have only 2x5 or 3x5 rafters on two foot centers and have easily withstood a century of time, although the rafters tend to be oak or some other strong hardwood, and are usually braced with collar ties and/or inside walls. And of course, the solid sheathing adds to the strength of the roof.

Modern building styles don't lend themselves well to roof construction intended for a slate covering. A current construction fad involves the use of 2x4 or 2x6 roof trusses on two foot centers. The trusses are pieces of pine or other softwood stapled together, and although popular with modern builders, they're not the best option for a slate roof. Better to "stick-build" the roof from solid, full size (e.g. 2x 8, 10, or 12) framing members nailed together. Ideally, these framing members, like the roof sheathing, would also be made from local lumber. Softwood trusses will work, however, as a last resort.

The author has designed and/or built many roofs, all of local (green) lumber, as was the style when most slate roofs were originally constructed at the turn of the last century. Building with full-sized, unplaned, rough sawn, green local lumber requires no special tools or equipment, and it allows for the construction of a solid, long-lasting roof. Again, the lumber shrinks and dries in place, causing no detriment to the structure or the roof. Instead, the structure becomes

Figure 12.6A
New "carriage house," designed and roofed by the author for the late Professor Robert L. Sayre, is built entirely of local hemlock, white oak, and maple. The local lumber was only a couple of weeks "off the stump." The four-gable (cross gable) roof is covered with recycled sea green slate, with purple slate decorations, and includes 20 ounce copper valleys. Slope is 10:12.
(Photo by author)

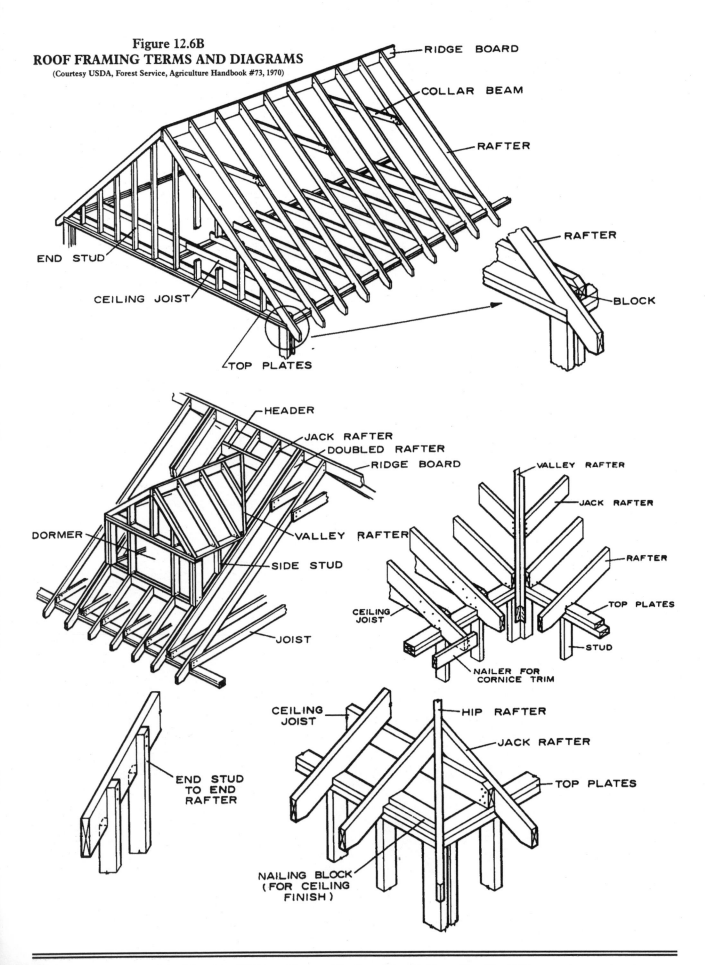

Figure 12.6B
ROOF FRAMING TERMS AND DIAGRAMS
(Courtesy USDA, Forest Service, Agriculture Handbook #73, 1970)

RIDGE BOARD

COLLAR BEAM

RAFTER

END STUD

CEILING JOIST

TOP PLATES

RAFTER

BLOCK

HEADER

JACK RAFTER

DOUBLED RAFTER

RIDGE BOARD

DORMER

VALLEY RAFTER

SIDE STUD

JOIST

VALLEY RAFTER

JACK RAFTER

RAFTER

TOP PLATES

CEILING JOIST

STUD

NAILER FOR CORNICE TRIM

END STUD TO END RAFTER

CEILING JOIST

HIP RAFTER

JACK RAFTER

TOP PLATES

NAILING BLOCK (FOR CEILING FINISH)

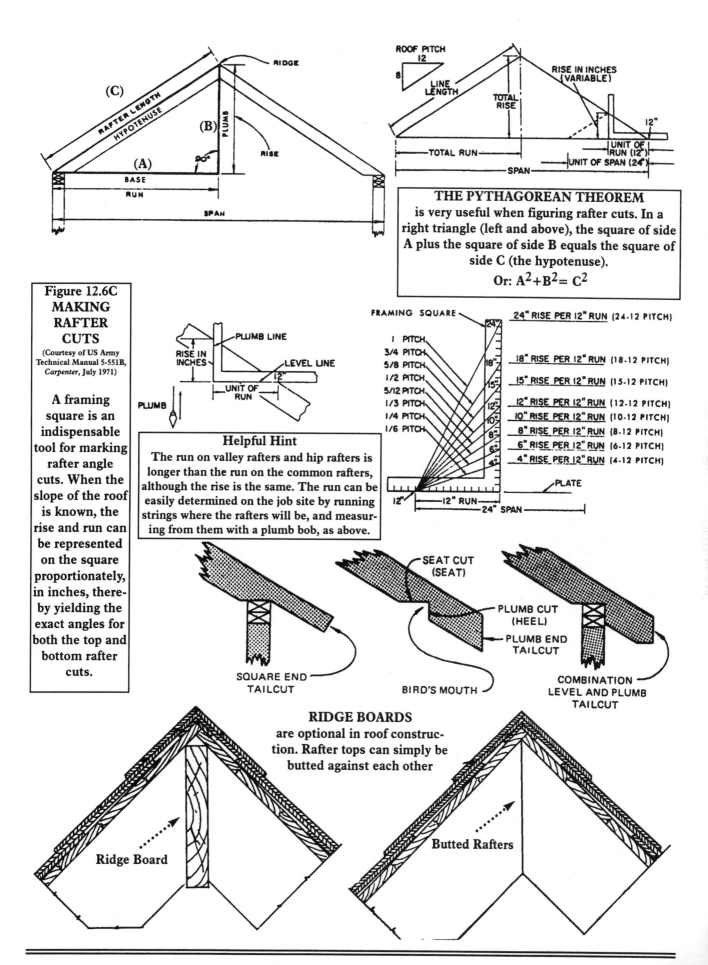

(C)

RAFTER LENGTH
HYPOTENUSE

RIDGE

(B)

PLUMB

RISE

(A)

90°

BASE

RUN

SPAN

ROOF PITCH
12
8
LINE
LENGTH

RISE IN INCHES
(VARIABLE)

TOTAL
RISE

12"

UNIT OF
RUN (12")

TOTAL RUN

UNIT OF SPAN (24")

SPAN

THE PYTHAGOREAN THEOREM

is very useful when figuring rafter cuts. In a right triangle (left and above), the square of side A plus the square of side B equals the square of side C (the hypotenuse).

$$Or: A^2 + B^2 = C^2$$

Figure 12.6C
MAKING RAFTER CUTS

(Courtesy of US Army Technical Manual 5-551B, *Carpenter*, July 1971)

A framing square is an indispensable tool for marking rafter angle cuts. When the slope of the roof is known, the rise and run can be represented on the square proportionately, in inches, thereby yielding the exact angles for both the top and bottom rafter cuts.

PLUMB LINE

RISE IN INCHES

LEVEL LINE

UNIT OF RUN

12"

PLUMB

Helpful Hint

The run on valley rafters and hip rafters is longer than the run on the common rafters, although the rise is the same. The run can be easily determined on the job site by running strings where the rafters will be, and measuring from them with a plumb bob, as above.

FRAMING SQUARE

1 PITCH
3/4 PITCH
5/8 PITCH
1/2 PITCH
5/12 PITCH
1/3 PITCH
1/4 PITCH
1/6 PITCH

24"
18"
15"
12"
10"
8"
6"
4"

12"

12" RUN

24" SPAN

PLATE

24" RISE PER 12" RUN (24-12 PITCH)

18" RISE PER 12" RUN (18-12 PITCH)

15" RISE PER 12" RUN (15-12 PITCH)

12" RISE PER 12" RUN (12-12 PITCH)

10" RISE PER 12" RUN (10-12 PITCH)

8" RISE PER 12" RUN (8-12 PITCH)

6" RISE PER 12" RUN (6-12 PITCH)

4" RISE PER 12" RUN (4-12 PITCH)

SQUARE END TAILCUT

SEAT CUT (SEAT)

BIRD'S MOUTH

PLUMB CUT (HEEL)

PLUMB END TAILCUT

COMBINATION LEVEL AND PLUMB TAILCUT

RIDGE BOARDS

are optional in roof construction. Rafter tops can simply be butted against each other

Ridge Board

Butted Rafters

Table 12.4

NUMBER OF SLATES AND NAILS FOR 100 SQUARE FEET OF ROOF

Size of Slate (inches)	Exposed Length (inches)	#of Slates (3 inch headlap)	Approx. Wt. Galv. nails (pounds)	Spacing of Lath
6x10	3 1/2	686	7 7/8	3 1/2
7x10	3 1/2	588	6 3/4	3 1/2
8x10	3 1/2	514	5 7/8	3 1/2
6x12	4 1/2	533	6	4 1/2
7x12	4 1/2	457	5 1/4	4 1/2
8x12	4 1/2	400	4 5/8	4 1/2
9x12	4 1/2	355	4 1/8	4 1/2
10x12	4 1/2	320	3 5/8	4 1/2
7x14	5 1/2	374	4 1/4	5 1/2
8x14	5 1/2	327	3 3/4	5 1/2
9x14	5 1/2	290	3 3/8	5 1/2
10x14	5 1/2	261	3	5 1/2
12x14	5 1/2	218	2 1/2	5 1/2
8x16	6 1/2	277	3 1/8	6 1/2
9x16	6 1/2	246	3	6 1/2
10x16	6 1/2	222	2 1/2	6 1/2
12x16	6 1/2	185	2 1/8	6 1/2
9x18	7 1/2	213	2 1/2	7 1/2
10x18	7 1/2	192	2 1/4	7 1/2
11x18	7 1/2	175	2	7 1/2
12x18	7 1/2	160	1 7/8	7 1/2
10x20	8 1/2	170	2 3/8	8 1/2
11x20	8 1/2	154	2 1/8	8 1/2
12x20	8 1/2	141	2	8 1/2
14x20	8 1/2	121	1 7/8	8 1/2
11x22	9 1/2	138	2	9 1/2
12x22	9 1/2	126	1 3/4	9 1/2
14x22	9 1/2	109	1 1/2	9 1/2
12x24	10 1/2	114	1 5/8	10 1/2
14x24	10 1/2	98	1 3/8	10 1/2

[Sources: Radford's Estimating and Contracting (1913), p. 252; and Slate Roofs (1926), p. 12.]

stronger as the lumber dries. This style of construction has been achieved on hundreds of thousands, if not millions of buildings across the USA. Most of the men who built local-lumber buildings are dead however, as are the slaters who roofed the buildings. Many of today's modern builders and architects have discarded the traditional building methods for ones that focus on low-cost and convenience. And now, unfortunately, it's difficult to find someone in the building trades who will support traditional building styles.

As we are beginning to see, there are a number of factors involved in determining the strength of the roof. First, the weight of the roofing must be taken into consideration. Secondly, the type of lumber must be noted, as some lumber species are much stronger than others. Thirdly, the size and spans of the framing members are important, as unsupported spans that are too long will sag over time. And lastly, the slope of the roof is important, as the lower the slope of the roof, the more snow weight it must bear, while the steeper the roof, the more wind-force it must withstand (see previous page).

The rule of thumb on slate roof slope is never to go below a 4:12 pitch (see Figures 12.6 and 12.6C). The lower the pitch, the longer the headlap, so that a 4:12 roof requires a 4 inch headlap. The steeper the roof, the shorter the headlap, so that a 12:12 roof requires only a two inch headlap (briefly jump ahead to Figure 12.11, page 201, to see "headlap" illustrated). Most slate roofs have a slope between 8:12 and 12:12, while 10:12 is perhaps the most common. Many of these older roofs have only a two inch headlap, despite the cautions of earlier roofing publications insisting that a three inch headlap is imperative. In fact, most older roofs have ignored many of the recommended conventions in slate roof construction (i.e. headlap, rafter size, flashing methods, etc.), and have fared quite well. Consequently, the information in this book reflects what's been recommended in print by earlier roofing publications, as well as what's been tried and proven.

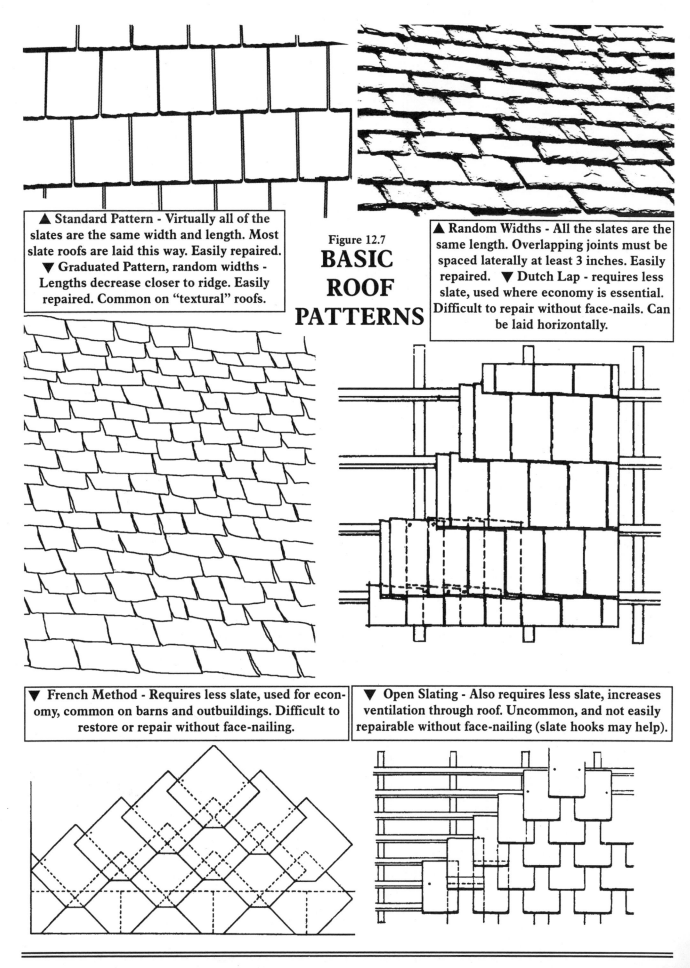

▲ Standard Pattern - Virtually all of the slates are the same width and length. Most slate roofs are laid this way. Easily repaired.
▼ Graduated Pattern, random widths - Lengths decrease closer to ridge. Easily repaired. Common on "textural" roofs.

Figure 12.7
BASIC ROOF PATTERNS

▲ Random Widths - All the slates are the same length. Overlapping joints must be spaced laterally at least 3 inches. Easily repaired. ▼ Dutch Lap - requires less slate, used where economy is essential. Difficult to repair without face-nails. Can be laid horizontally.

▼ French Method - Requires less slate, used for economy, common on barns and outbuildings. Difficult to restore or repair without face-nailing.

▼ Open Slating - Also requires less slate, increases ventilation through roof. Uncommon, and not easily repairable without face-nailing (slate hooks may help).

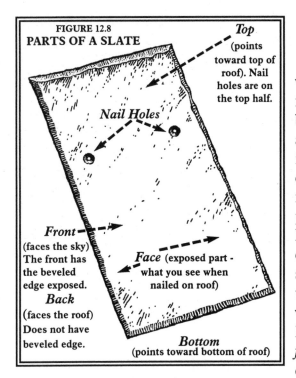

FIGURE 12.8
PARTS OF A SLATE

Top (points toward top of roof). Nail holes are on the top half.

Nail Holes

Front (faces the sky) The front has the beveled edge exposed.

Back (faces the roof) Does not have beveled edge.

Face (exposed part - what you see when nailed on roof)

Bottom (points toward bottom of roof)

When nailed to a roof, slates must be nailed with their top facing the top of the roof, and their front facing out (Figure 12.8). This may seem elementary, but slates do have a top and bottom, front and back, and they must be oriented properly on the roof. In rare instances slates *can* be flipped over and laid back side out, when they're (for example) being recycled from a roof that has been painted or coated, so as to hide the paint, and they will function quite well this way. However, they won't look right because the beveled edges will be hidden along with the front side of the slate, and it's the bevels that give the edges of the slate their distinctive look. When slates are placed backwards the nail holes must be re-punched, as the nail holes are beveled so as to accept the nail head on the *front* of the slate. In general, always lay the slate face out and top up. Furthermore, butt the slates against each other on the sides. There is no advantage to leaving a space between the slates. The slates will not expand and contract and push against each other as wood does.

Most older slate roofs in America are laid in a standard pattern consisting of slates that are, on each roof, virtually all the same width and length (which varies from roof to roof - see Table 12.4), and thickness (about 3/16"). A rule of thumb is to use larger slates on larger roofs, and smaller slates on smaller ones (this rule is frequently broken). Larger slates go on faster, as fewer are required to cover a given area. The largest standard size is 14" wide x 24" long, and only 98 of these will cover one hundred square feet of roof surface with a three inch headlap, using about 200 nails. An 8x10 slate, on the other hand, requires 514 slates to cover the same area, and about 1,050 nails. Obviously then, the larger slates require less time and labor to install.

Some styles of laying slate were developed in order to conserve materials, such as the Dutch Lap, Open Slating, and French methods (Figure 12.7), which require far fewer slates to cover a roof than the standard overlap pattern does. These styles tend to be found on barns and

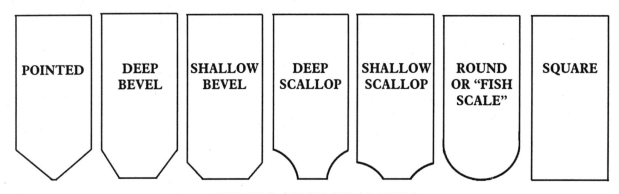

| POINTED | DEEP BEVEL | SHALLOW BEVEL | DEEP SCALLOP | SHALLOW SCALLOP | ROUND OR "FISH SCALE" | SQUARE |

Figure 12.9   **SHAPES OF ROOF SLATES**
Roof slates were once manufactured in decorative shapes in order to add an elegant appearance to the roof. Today's roof slates are manufactured in the rectangular shape (above right), however they can still be cut to match the patterns of old roofs using only a simple hand cutter (pages 172-173).

outbuildings where the property owner didn't want to spend any extra money on materials. The main problem with these conservative slating styles is that the roofs are consequently difficult to restore properly when they get old.

Figure 12.10
Roof jacks and planks make a handy and safe platform from which to work when installing a new slate roof. The roof above is made of recycled sea green slate. (Photo by Barry Smith)

The standard lap style is ingenious in that it allows for easy replacement of broken slate in such a manner that the nail holding the replacement slate is covered by flashing and therefore rendered leakproof. The side-lapped slates do not include this option in their design, and therefore when replacing them, one is almost limited to face-nailing the slate, which is not good for the roof in the long run. Therefore, when installing a slate roof, it's recommended that the slates be laid in a standard lap fashion.

When most of the American slate roofs were installed a century ago, it was customary to use slates that were all the same width, length and thickness on one roof. Some roofs had 12"x22" slates on them, some had 9"x18", etc., and the slates were the standard 3/16 inch thick. The exception to this rule, however, involved using a "graduated" design, which utilized longer (and usually thicker) slates near the bottom of the roof, gradually decreasing in length and thickness nearer the top - a design common in Wales, which tended to be limited to the more affluent homes in America. Another roof style was the "textural" roof, which simply utilized slates of various widths, lengths, colors, and thicknesses, creating a mottled looking roof with a highly aesthetic appeal; again reminiscent of the old country, and also limited to the more expensive homes and buildings in the USA.

Today it's common to see new slate roofs laid with a random width pattern - the slates are all the same length, but random widths. Random width slates are less expensive than uniform width slates, and they work just as well, so they're preferred by some roofers. The joints, or slots

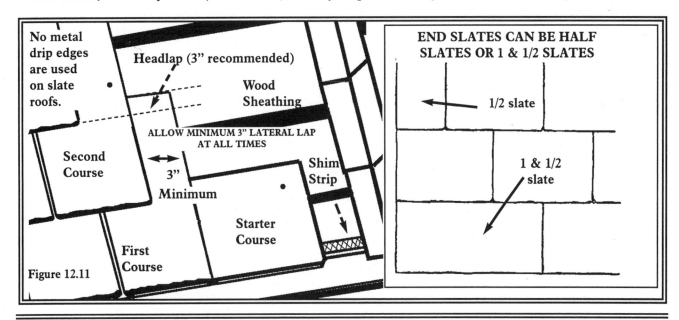

No metal drip edges are used on slate roofs.

Headlap (3" recommended)

Wood Sheathing

ALLOW MINIMUM 3" LATERAL LAP AT ALL TIMES

Second Course

3" Minimum

Shim Strip

First Course

Starter Course

Figure 12.11

END SLATES CAN BE HALF SLATES OR 1 & 1/2 SLATES

1/2 slate

1 & 1/2 slate

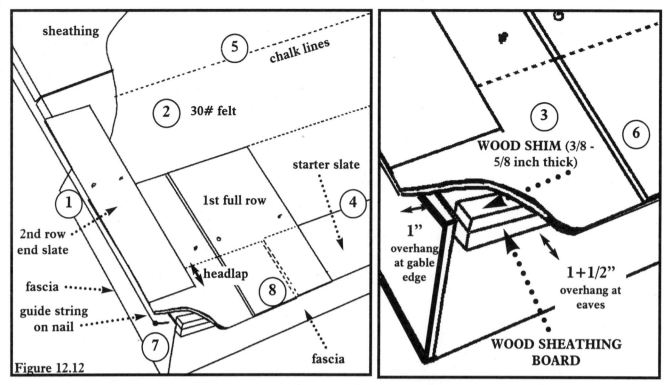

Figure 12.12

When Starting to Slate a Roof - Ten Quick-Reference Steps

1) Make sure that the fascia is completely installed beforehand and that the ends of the sheathing boards are firmly nailed. 2) Felt over the solid one inch thick local-lumber sheathing with 30 lb. roofing felt, lapped at least three inches at the top. 3) Nail a wooden starter shim at the bottom edge of the lowest sheathing board - it should be 1/2" to 5/8" thick, and at least an inch wide (eight foot lengths are convenient). Cedar or redwood is ideal (cedar shim shingles will work), but the same local lumber as the sheathing will do. 4) Chalk a horizontal line on the felt paper for the starter slates, measuring the <u>width</u> of the slate up the roof from the bottom edge of the wood shim, deducting 1+1/2 inches for the slate overhang. Then chalk a line for the first full row, now measuring up the roof the <u>length</u> of the slate and deducting an inch and a half for the overhang. 5) Now measure up the remainder of the roof equal distances equivalent to the *exposure* of the slate, and chalk lines accordingly. But first, make sure your second full row of slates will overlap the starter row by three inches based on your measurements - if not, drop that second row down to where you need it to be, *then* chalk the rest of the roof with the exposure measurement. *[Exposure is determined by subtracting the headlap from the total length of the slate, then dividing the remainder in half. For example, a 20" slate with a 2" headlap will have a 9" exposure (20 - 2 = 18, divided in half = 9). Or subtract 3" from the length and divide the remainder in half to determine the exposure when using the recommended 3" headlap.]* 6) Do not bed the starter slates or any slates in roof cement or caulk. Adhesives make it very difficult to repair the roof in the future. Instead, two 1+1/2" hot dipped galvanized or copper nails per slate is a good rule of thumb which will ensure the secure attachment of all slates to the roof. Don't nail the slates *too* tightly, let them hang on the roof. Do make sure the nailheads are set into the slate however, as nails that stick up will eventually wear a hole in the overlying slate, and cause a leak. 7) Tap a couple of temporary nails into the side of the fascia on the gable end, one at the top and one at the bottom, and run a string up the edge of the roof positioned one inch out from the fascia. Use the string as a guide to align the edge of the slate as you nail them into place. Remove the string when you're done. 8) Make sure the slots between the slates on the <u>first</u> <u>full</u> row are staggered at least 3" laterally from the butted ends of the <u>starter</u> slates. If not, reinforce the joint by sliding a piece of flashing over the starter slate and under the first row at the joint. 9) You can work the first half dozen rows from a ground ladder or ground scaffold, then nail roof jacks and planks along the bottom of the roof and work up from there. Use more jacks and planks as needed. 10) Have fun!

> *Exposure is determined by subtracting the headlap from the total length of the slate, then dividing the remainder in half. A 20" slate with a 2" headlap will have a 9" exposure (20 - 2 = 18, divided in half = 9). Or subtract 3" from the length and divide the remainder in half to determine the exposure when using the recommended 3" headlap.*

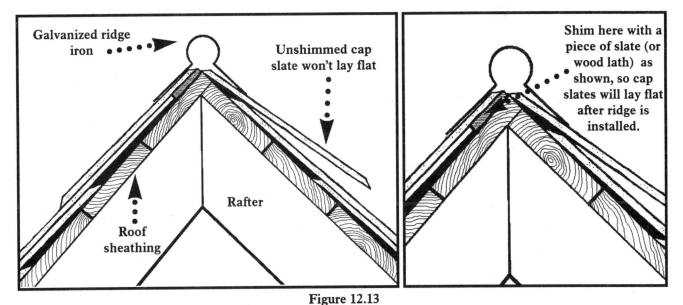

Figure 12.13
CAP SLATES OFTEN NEED SHIMMED IN ORDER TO LAY FLAT
Use a piece of slate (or wood) to shim the upper edge of the cap slates as needed to prevent them from being cocked up off the surface of the roof (as shown above, on the right side of the roof). The left side has been shimmed, so the slates lay flat.

between the slates must have at least 3" of lateral spacing where they overlap the course below, and this little bit of extra critical detail when laying the slate increases the overall labor time involved when installing a random width roof.

When laying out a roof in preparation for slating it, chalk lines across the entire roof area marking the *tops* of each row of slate. *No metal drip edges are needed on slate roofs* and they serve no useful function, so don't bother nailing any on before slating. When measuring for the starter slate and the first row, allow for the slate to hang beyond the drip edge of the fascia (or trim moulding) one and a half inches. The starter slates are usually made of the same size slates as those on the main roof, turned sideways, and usually 1/4 of the length of the first one is trimmed off to allow the joints to be properly staggered in relation to the overlapping row. *The rule of thumb is that all butt joints between slates should have a minimum of three inches of lateral clearance in relation to the butt joints of overlapping slates.* On many old roofs the starter slates are not laid sideways, but are simply the same slates as the rest of the roof - cut short - and again the joints are staggered. In all cases, the starter slate must be laid over a shim strip about 1/2 inch thick, which cocks the slate at an angle comparable to the angle of the slates on the rest of the roof (Figure 12.12).

The slates that run up the side of the roof should extend beyond the gable ends one full inch. Run a string up the edges of the roof to give yourself a straight edge to follow when laying the slate (tie the string to temporary nails).

When you reach the top of the roof, the top two rows of slates must be cut shorter in length to fit the roof. Frequently the top row, the "cap" slates, must be shimmed so they'll remain flat when the ridge iron is installed; otherwise they'll cock crookedly and look bad. They can be shimmed with pieces of slate, usually the pieces that are cut off the top rows when the slate is laid. (Figure 12.13).

Figure 12.14
VALLEYS
(Modified from <u>Slate Roofs</u> (1926), author unknown, pp. 20-21)

The standard valley is left open with about six inches exposed, has no center crimp, and is nailed along the edges; total width ranges from about 14" to about 20", with 16" being recommended. Length of sections should not exceed 12', with 10' maximum recommended. Sections are overlapped six inches. No roof cement or adhesives are used under or over valleys or slate. Slate edges run parallel to each other, although some roofers prefer the exposed valley to widen toward the bottom. Cleated valleys (below, right) are preferred by some roofers. The standing "V" crimp shown below, is used when a steep roof pitch drains onto a shallow one - the "V" prevents water from forcing its way under the slates on the shallow side. A center straight-line crimp is needed in closed or very narrow valleys, or can be added simply for style.

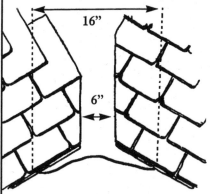

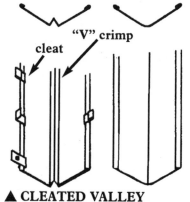

▲ STANDARD OPEN VALLEY
Not cleated, no center crease

▲ STANDARD OPEN VALLEY
Cross Section

▲ CLEATED VALLEY
"V" Crimp (left), Center crease (right)

CLOSED VALLEYS

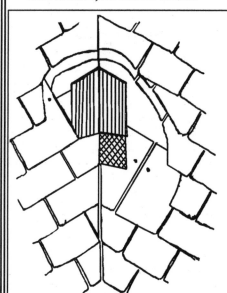

show no metal in the valley, since the slates butt against each other in the valley center. They can be installed in two basic ways: with continuous flashing like standard valleys (shown right), or with step flashing (shown left). Continuous closed valleys should have a center crease, and should be ten inches wide. No center V crimp is necessary. Stepped closed valleys follow the standard rule of thumb when using step flashing: *The bottom of the step flashing lines up with the bottom of the slate on top of it, the top of the step flashing lines up with the top of the slate underneath it.* (The top of the step flashing may extend above the underlying slate, for nailing purposes.)

10 inches of metal

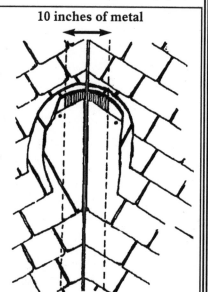

QUICK REFERENCE - VALLEY INSTALLATION

Make sure roof sheathing is covered with 30 lb. felt paper. Run a chalk line up one side of the valley to indicate where the flashing should be positioned (for 16" flashing measure 8" from valley center). Use 16 or 20 ounce copper, stainless steel (18 gauge), or heavy aluminum (.040"), 16 inches wide (10" for closed valley). Valley flashing does not need to be cleated, crimped, or creased, but should have a V crimp if a steep roof is running into a shallow one; valley should have a center crease if the valley is especially narrow and on a steep roof (or closed). Lay valley into roof, carefully forcing it into place using your knee, and nailing within 1" of edges, using nails of compatible metal. Keep metal valley sections to a maximum of 10', overlap 6" at ends, don't use roof cement or adhesives. Chalk lines up valley flashing to indicate edges of slates (3" out from center for 6" exposed valley), then immediately mark over lines with a permanent felt-tipped marker. Leave six inches exposed for standard open valley, chalk lines at an angle for open valley that widens near bottom, if desired. After valley metal is in place, chalk remainder of roof for slates, chalking into valley. When nailing slates over valley, hold nails back to within an inch of the edge of the metal.

VALLEYS

Valley metal flashing is installed over the felted sheathing before any slating begins. The felt paper need not overlap the valley flashing (the valley metal can be laid right on top of the felt). Before the metal is installed, a chalk line is struck up the edge of one side of the valley, on the felt paper, to indicate where the edge of the metal valley flashing should be. The metal is then nailed in place with a nail of a compatible material (i.e. copper flashing with copper or brass nails, etc.), and the nails are kept to within one inch of the edge of the valley metal.

Valleys don't need to be creased on a brake, leaving them with a bend line down the center, except in cases where the two roof surfaces are very steep and the valley is very narrow. Otherwise, a creased valley is simply a matter of personal style and creates no advantage in the function of the valley. Instead, the valley metal is nailed along one edge, then carefully forced into the roof with the pressure of a knee as the other side is nailed.

Valleys should be laid in sections not to exceed twelve feet in length, although a ten foot maximum length is recommended due to the adverse effect of expansion and contraction that can cause long pieces of metal to buckle and develop a leak over time. The valley sections should simply be overlapped by six inches - no soldering is necessary or recommended, as it's the old solder joints on the old valleys that tend to leak first, once again, due to expansion and contraction. Do not use roof cement or other adhesives along the edges of a valley, as these make later repairs of the roof unpleasant and difficult while adding no advantage to the functioning of the roof.

Furthermore, valleys do not need to be cleated along the edges, that is, fastened with bent pieces of metal called cleats instead of nails (Figure 12.14). It was once recommended that valleys be cleated in place for fear of expansion and contraction wearing any nails loose along the edges, but most valleys were simply nailed in place anyway, and time has shown that this procedure works well.

Valleys are typically laid "open," with approximately six inches of metal exposed, while the overall width of the metal can vary from 14" to 20", although 16" is recommended (providing no nails penetrate the valley more than an inch or so from the edge of the metal when the roof is slated). Open valleys typically have parallel sides running from bottom to top, although some roofers prefer open valleys that gradually widen toward the bottom. When laying

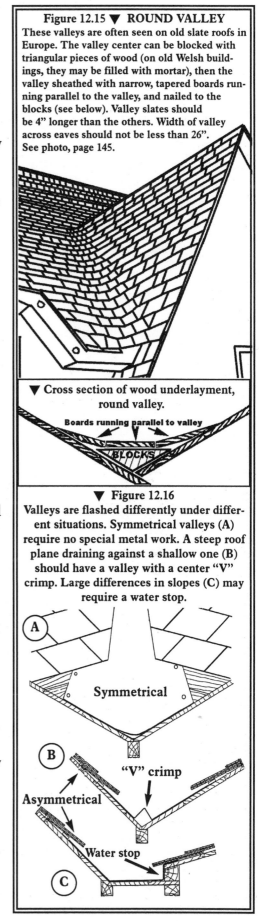

Figure 12.15 ▼ ROUND VALLEY
These valleys are often seen on old slate roofs in Europe. The valley center can be blocked with triangular pieces of wood (on old Welsh buildings, they may be filled with mortar), then the valley sheathed with narrow, tapered boards running parallel to the valley, and nailed to the blocks (see below). Valley slates should be 4" longer than the others. Width of valley across eaves should not be less than 26". See photo, page 145.

▼ Cross section of wood underlayment, round valley.

Boards running parallel to valley

BLOCKS

▼ Figure 12.16
Valleys are flashed differently under different situations. Symmetrical valleys (A) require no special metal work. A steep roof plane draining against a shallow one (B) should have a valley with a center "V" crimp. Large differences in slopes (C) may require a water stop.

A

Symmetrical

B

"V" crimp

Asymmetrical

Water stop

C

Figure 12.17
▲ **CLOSED VALLEY** - Slates are butted against each other in the middle of the valley.
(Photo by Adam Bossert)

Figure 12.18
▶ **VALLEY COPPER** being rolled out into a replacement valley on the barn shown on page 189. The copper is not crimped, creased, or cleated, just carefully force-fit into the valley, and nailed with copper nails. Each ten foot section is overlapped six inches.
(Photo by the author)

Figure 12.19
MITERED HIPS on a 90 year old Vermont purple slate roof, in Grove City, PA, are being covered with new, painted, 26 gauge galvanized ridge iron. The hips had opened enough over the years to cause slight leaking, and therefore had to be covered.
(Photo by the author)

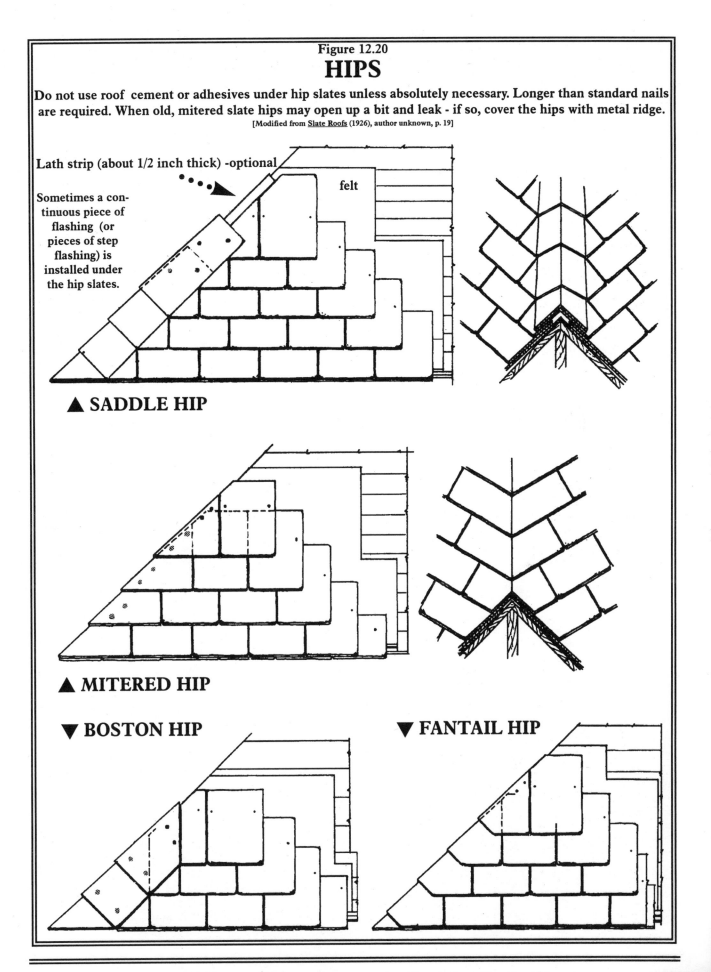

Figure 12.20
HIPS

Do not use roof cement or adhesives under hip slates unless absolutely necessary. Longer than standard nails are required. When old, mitered slate hips may open up a bit and leak - if so, cover the hips with metal ridge.

[Modified from <u>Slate Roofs</u> (1926), author unknown, p. 19]

Lath strip (about 1/2 inch thick) -optional

felt

Sometimes a continuous piece of flashing (or pieces of step flashing) is installed under the hip slates.

▲ SADDLE HIP

▲ MITERED HIP

▼ BOSTON HIP

▼ FANTAIL HIP

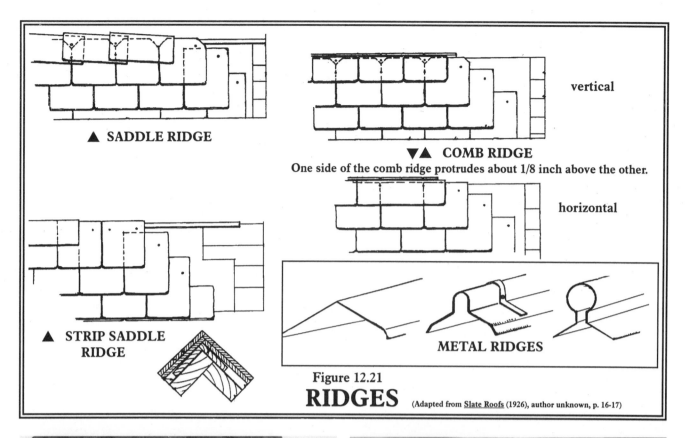

▲ SADDLE RIDGE

vertical

▼▲ COMB RIDGE

One side of the comb ridge protrudes about 1/8 inch above the other.

horizontal

▲ STRIP SADDLE
RIDGE

METAL RIDGES

Figure 12.21
RIDGES

(Adapted from Slate Roofs (1926), author unknown, p. 16-17)

Figure 12.22 (Left) Lead ridge common in Great Britain, here shown in Devon, England. (Photo by Dave Starkie).
(Right) Copper ridge in Massachusetts by Doug Raboin. (Photo by Doug Raboin)

slate into a valley, chalk a line the length of the valley on both sides to indicate the edges of the slate, then draw over the chalk lines with a permanent ink felt-tipped pen, as the chalk lines will wear off the metal almost immediately. When nailing slate over the valley metal, be careful to nail only along the edge of the metal, and not anywhere near the center. If a small, triangular piece of slate cannot be nailed over the valley at the end of a row without nailing too close to the center, eliminate that piece of slate - you won't need it.

Valleys can also be laid "closed," with no metal exposed (Figures 12.14 and 12.17), and "rounded" - a style more common in Wales and Europe, in which no metal flashing need be used at all (Figure12.15). Refer to Chapter 14 (Flashing) for more information about valleys.

HIPS

Many hips on slate roofs are simply covered with metal ridge flashing, either galvanized steel or copper, but usually galvanized steel ("ridge iron"). These are simply nailed into place using eight penny galvanized or common nails spaced every foot or two, and the nail heads are caulked with a material like clear silicon of a 50 year durability.

Hips made of slate are quite common and require less maintenance than metal hips, which usually must be kept painted. Common slate hips include the miter hip, and the saddle hip (Figure 12.20). Mitered hips simply consist of the standard roof slates butted against each other at the hip. In some cases, mitered hips have step flashings under the hip slates. Saddle hips are made of slates positioned parallel to the hips over the standard roof slates, and in some cases may have flashing underneath them, usually a continuous piece rather than step flashing. Flashing under the hips reinforces the hips and is helpful in preventing leakage at the hip when the roof ages, since the hip slates tend to separate somewhat after eighty or ninety years. Nevertheless, flashing under hip slates is uncommon. Instead, ridge iron can be installed over old, separated hips as in Figure 12.19, by nailing it in place with eight penny galvanized nails, caulked on the nail heads.

Figure 12.23
Lead hips and valleys plus ornate ridges on this slate roof in Worcester, England, demonstrate the use of durable materials to finish off a roof.
(Photo by Dave Starkie)

Figure 12.24
Exposed rafters on a Welsh
slate roof. The old slates and
lath have been removed by
Robert Jones and David
Hussey in preparation for re-
roofing.

Figure 12.25
The felt is installed directly
over the rafters.

Figure 12.26
The slating lath is nailed to the
rafters directly over the felt
paper.
(Photos this page by the author)

RIDGES

Ridges, like hips, are typically made of metal (in the United States), especially galvanized steel. When finishing slating along a ridge, it's important that the roof sheathing does not have any appreciable gap at the peak. If a gap exists (as is typically left when a carpenter sheaths a roof for ventilated ridge) the slates will not lay properly, and the ridge iron will not have a sufficient base in which to nail. Furthermore, the cap slates will sometimes not lay flat when the ridge iron is nailed into place, so they must be shimmed beforehand as in Figure 12.13. A piece of wood lath may also serve as a shim behind cap slates as well as behind ridge slates if needed. The Welsh are quite fond of lead ridges, unavailable in the United States, but certainly superior to galvanized steel ridge (Figure 12.22). Copper ridge is also available in the US, either hand-made as in Figure 12.22, or prefabricated from roofing supply outlets.

Slate ridges are quite common in the US, and some are illustrated in Figure 12.21. The Welsh prefer ceramic tile ridges on their slate roofs (Figure 12.23), as can be seen throughout this book; these ridges are perhaps the most durable of all. They're more compatible with the masonry-walled construction of Great Britain, and they're almost unheard of in the United States.

THE WELSH STYLE

As stated earlier in this chapter, the Welsh originally hung their slate on split wooden lath using wooden pegs as hangers. The pencil-stub sized pegs were split square out of a block of wood, then driven through a round hole that had been punched into the top center of the slate. This force-fitting caused the peg to be firmly wedged in the slate, and the entire system worked very well and lasted quite a long time.

The system was modified over the years, however, so that the slates were eventually nailed into the lath - the lath itself no longer split, but sawed into approximately one inch by two inch strips. Felt paper is installed *under* the lath where it drapes between the roof rafters waiting to catch a leak should any occur. The advantage to felting under the lath is that the roof nails do not penetrate the felt. When a roof of this style gets old enough to need to be replaced, the slates are carefully removed, the lath and felt are replaced with new material, then the roof is reslated, either with new slates or old.

The slates are sorted by thickness before being nailed to the roof. The "very heavies" are placed at the bottom of the roof, the "heavies" above them, the "mediums" above the heavies, and the "lights" at the top. Unlike American slates, the Welsh slates don't come pre-punched for nail holes from the quarry, but instead must be "holed" by the roofer prior to installation. Figures 12.24 through 12.27 illustrate the Welsh roofing process.

Another modern style in Welsh slate roofs is "weather-clipping" the slates at the gable ends of the roof (known as the "gable verge" in Wales). This involves cutting the outside corner off the bottom of the edge slates to draw the rain water toward the center of the roof as it runs down the "verge" (Figure 12.28).

RECYCLING ROOF SLATE

One of the unfortunate consequences of America's throw-away mentality is the loss of thousands of perfectly good slate roofs, which are ripped off and destroyed by uncaring roofers who can't be bothered with recycling anything. The slates are dumped in landfills, then the roof

Figure 12.27
Twenty-seven year old Neil Berridge, with ten years of slating experience under his belt, prepares this new life-boat shed in Barmouth, Wales, for slate. The roof is papered and the lath is installed. The slates have been sorted into "very heavies," "heavies," "mediums," and "lights," depending on thickness - the heavier ones go at the bottom of the roof. The slates have also been "holed" by Neil (nail holes punched). The finished roof is shown below. (Photos by author)

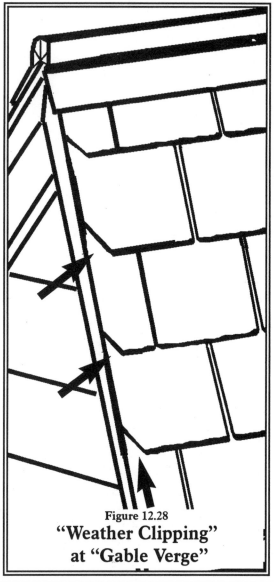

Figure 12.28
"Weather Clipping" at "Gable Verge"

Figure 12.29
The huge barn near West Middlesex, PA, shown above left, was scheduled for demolition so its roof slates were salvaged . Architect Chris Leininger, left, pries loose the slates, which slide down to a waiting plank (as illustrated in the bottom photo of another roof being salvaged), then they're slid down a chute into the barn, landing on an old mattress (above). Finally, the slates are taken to Pennsylvania's Slippery Rock University Harmony Homestead and installed on a new building, shown at left with some of the student construction workers.

(Photos by the author)

Figure 12.30
A tornado flattened this Mercer, PA garage (top), but the slate roof was salvaged by astute owners Mike and Diane Sharr, and when the new garage was built (center), the slates went back on.
The Hooker, PA roof at the bottom, designed by Guido Lesser, displays a clever mix of sizes, shapes and colors, yielding an aesthetically unique recycled slate roof.

(Top photo by Mike Sharr, center photo by the author, bottom photo by Guido Lesser)

Figure 12.31
All of these roofs are made of recycled slate, and should last several human generations, if not a century or more. They include a writer's residence (above), a professor's home (right), a beach house/sauna (below left), and a poet's retreat (below right).
(Photos by the author)

is replaced with disposable petro-chemical roofing which is destined to continue clogging land-fills indefinitely. Most hard slate roofs are quite recyclable, however, and care should be taken when removing them to salvage the slates, either for slate roof repairs or for a completely new roof.

The slates should be pried loose with a flat pry bar ("wonderbar") starting at the top of the roof, and then allowed to slide down the roof and collect on planks run across the bottom of the roof on roof jacks. While someone is prying the slates loose, someone else can be gathering them up and either carrying them down a ladder, dropping them down over the eaves or through a hole in the roof in a rope and harness, or sliding them down a chute to someone else. Ideally, they would be slid directly into the back of a waiting truck where they'll be carefully stacked on edge before transit.

Some of the slates will break during the process of removal, but it's better that they break there than after they've been nailed onto another roof. Some slates will develop hairline cracks or other flaws and must be discarded. A good slater can tell a bad slate by simply holding it in his hand and tapping it. A bad slate will give a dull thud, a good slate will ring. When removing slates from an old roof, it's better to pry each nail loose than to pull the nails through the slates. Prying the nails out preserves the old nail holes, while pulling the nails through the slates ruins the holes which must then be re-punched with a slate hammer before the slate can be nailed to another roof (unless the slate is used for repairs where the holes may not be used at all).

Recycling hard slate roofs is a good way to get an excellent roof for a new building with an inimitable antique look that will last a century. Often the recycled slates from one building don't yield enough quantity to cover another roof, and recycled slates from two or more roofs must be collected before a sufficient quantity is obtained for the job. When this situation occurs, it is imperative to "shuffle" the slates together before nailing them onto the new roof, as each old roof over time has developed its own weathered appearance, and in order for the recycled slate roof to look right the slates must be randomly mixed. For example, if you remove a thousand slates from old roof "A," and five hundred from old roof "B," and you need fifteen hundred for the new roof you're going to install, then for every two "A" slates you carry up onto the new roof, you must carry up one "B" slate. If you can carry twelve slates up a ladder at a time by hand, then eight of

Figure 12.32 Recycled slates cover the roofs of this dwelling as well as the back wall of the sun-space, where the slate's ability to absorb heat has aided the growth of this January lettuce, an unusual Pennsylvania crop.
Photo by the author.

them should be "A" slates and four of them "B" slates. This is how it must be done if you want it to look right.

Alternatively, if you have collected slates from two dramatically contrasting roofs, such as from a green roof and from a purple roof, then you may want to consider designing a pattern into the roof to take advantage of the color contrast. A good example of this is shown on the garage with the diamond patterned roof made from sea green slate and purple slate illustrated in the center color photo section of this book.

Putting designs into roofs is a fun thing to do, but in order to do it properly one must make a schematic diagram of the roof beforehand, and draw the design onto it. Then the schematic is taken up onto the roof when the slates are installed and referred to by the roofer as the slate is laid. A schematic is simply a line drawing showing the rows of slate in their proper proportions. The exact size of the slate, as well as the exposure, must be known before a schematic can be accurately drawn.

It's often a good idea to increase the headlap on recycled slates when nailing them to a new roof in order to cover up the weather marks that remain on the old slates. If the roof that the slates were removed from had a two inch headlap, the new roof should have a two and a half inch headlap, or even a three inch headlap. This will give the finished recycled-slate roof job a cleaner appearance. This extra lapping sometimes poses a problem, however, as the nail holes may then become too low and the nails will penetrate the top of the slates in the underlying row, which should be avoided if possible. If that happens you should poke new holes in the slate, higher up, before nailing.

OTHER USES FOR RECYCLED ROOF SLATES

Slates can also be recycled for purposes other than roofing. Painting and decoupage on old roof slate is popular among craftspeople, for example. Slates can also be cut up and stacked to make sculptures, or assembled into such things as doll houses, candle sticks, or vases with the proper adhesives *[Akemi Adhesives, Wood and Stone Co., 10115 Residency Road, Manassas, Virginia, 22110; ph: (703) 369-1236]*. Roof slates can be engraved with stone engraving chisels, or lettering and designs can be sandblasted into them using a sandblast stencil *[Design Services, Inc., PO Box 1789, Land O' Lakes, FL 34639; ph: 1-800-624-0075 or fax (813) 996-4523]*.

Old roof slates that are good and solid make a good floor covering too, especially on concrete surfaces. They can be laid directly onto the wet concrete, perhaps with a bonding agent painted to their underside, or glued to cured concrete or plywood with an epoxy (thin-set) tile adhesive, or even with trowel grade roof cement (allow a few weeks drying time when using roof cement as an adhesive). Roof slates can be walked on when laid flat over an unyielding surface, but should not be walked on when on a roof where they're overlapping each other and therefore likely to break. It's imperative that the floor have no "give" to it when using old roof slate as a floor covering. Slates can also be glued to drywall or plywood walls to make a very interesting and beautiful wall surface simulating cut stone in appearance. Again, epoxy tile cement is best, and with a cutter, a variety of slates, and an imagination, the design possibilities are endless.

Finally, roof slates can be epoxied to brick surfaces such as old chimneys to make them appear to be stone. In all cases, when recycling roof slate for decorative purposes, make sure the slate is not soft and flaking (use only good, hard slate), and thoroughly clean each slate with soap and water before use. And remember, if you cut the slate with a hand-operated slate cutter, the edges will be beveled. If you want square edges, you'll have to cut the slate with a masonry blade or a diamond blade on a circular saw or a grinding tool.

SOURCES OF NEW AND USED ROOF SLATE AND ACCESSORIES*

Buckingham-Virginia Slate Corporation, One Main Street P.O. Box 8, Arvonia, VA 23004-0008; Ph: (804) 581-1131, Fax: (804) 581-1130.
New Virginia slate and accessories.

Cwt y Bugail Slate Quarries Co. Ltd., Blaenau Ffestiniog, Gwynedd LL41 3RG, Wales UK; Ph: (0766) 830204, Fax: (0766) 831105.
New black Welsh roof slate.

David Camara and Sons, Route 22A, Hampton, NY 12837; Ph: (518) 282-9646; Fax: (518) 282-9906.
New and used roof slate, tools.

Anthony Dally and Sons, Inc., Pen Argyl, PA 18072; Ph: (610) 863-4172; Fax: (610) 863-8388.
New Pennsylvania blue-black roof slate.

Evergreen Slate Co., Inc., Granville, NY 12832; Ph: (518) 642-2530.
New roof slate, slate tools (cutters, hammers, rippers, hooks), Video on The Art of Handcrafting Slate.

Glendyne Quarry, 396 rue Principale, St-Marc du Lac Long, Quebec, Canada G0L 1T0; Phone: (418) 893-7221, Fax: (418) 893-7346.
Canadian slate similar to Monson slate, available through Newfoundland Slate Co.

Hilltop Slate, Inc., P.O. Box 201, Middle Granville, New York 12849; Ph: (518) 642-2270/642-1453; Fax: (518) 642-1220.
New roof slate and supplies.

Kennedy Slate Mine, Box 249, Monson, ME 04464; Ph: (207) 876-3761/ 876-2269.
Monson slate blocks.

The New England Slate Company, Burr Pond Road, Sudbury, VT 05733; Ph: (802) 247-8809; Fax: (802) 247-0089.
New and used roof slate, slate tools and slate bag.

Newfoundland Slate Inc., Sales: 8800 Sheppard Ave. E., Scarborough, Ontario, Canada M1B5R4, Ph: (416) 281-8181 or 1-800-975-2835, Fax: (416) 281-8842.
New Canadian roof slates and tools. Retailer for Trinity Slate Co. of Newfoundland, and Glendyne Slate Co. of Quebec.

Penn Big Bed, P. O. Box 184, Slatington, PA 18080; Ph: (610) 767-4601; Fax: (610) 767-9252.
New PA black roof slate, slate tools, hooks, nails, snow guards.

Penrhyn Quarries Ltd., Bethesda, Gwynedd, Wales UK.
New Welsh slate of various colors.

Renaissance Roofing, Inc., PO Box 5024, Rockford, Illinois 61125; Phone: (815) 874-5695; Fax: (815) 874-2957.
Source of new and used slates, specializing in used ceramic roof tiles.

Structural Slate Co., 222 East Main St., Pen Argyl, PA 18072; Ph: (610) 863-4141 or (800) 677-5283; Fax: (610) 863-7016.
New Pennsylvania roof slate, tools.

Trinity Slate, General Delivery, Burgoyne's Cove, Newfoundland, Canada A0C 1G0; Phone: (709) 663-4104, Fax: (709) 663-4105.
Roof slate almost identical to Welsh slate. Can be purchased through Newfoundland Slate (see above).

T.U.F.S. (James E. Kelly), 1820 Mill Pond Road, Wells, VT 05774; Ph: (802) 645-0010.
Hand forged slate working chisels and tools.

U.S. Quarried Slate Products, Inc., Scotch Hill Road, P.O. Box 261, Fair Haven, VT 05743; Ph: (802) 265-2029, 265-2030; Fax: (802) 265-3224.
New roof slate.

Vermont Slate and Copper Services, Inc.; 55-C Gonyeau Rd., Milton, VT 05468; Ph: (201) 848-8898, (802) 893-7703.
Weather vanes, finials, snow guards, new and used slate, consultation.

Vermont Structural Slate Co., Inc., Box 98, Fair Haven, VT 05743; Ph: (802) 265-4933/34 or (800) 343-1900; Fax: (802) 265-3865.
New roof slate, tools.

Virginia Slate Company, 100 East Main Street, Richmond, VA 23219; Phone: (804) 282-7929, or 1-888-VA SLATE; Fax: (804) 285-4442; email: VCA11@aol.com
Source of Buckingham County (black) slate.

Welsh Slate Ltd, Business Design Centre, Unit 205, 52 Upper St., London N1 0QH England; Phone: +44(171)354-0306; Fax: +44(171)354-8485; email: nblager@welshslate.com
Source of Welsh slate from the various Welsh quarries.

Williams and Sons, 6596 Sullivan Trail, Wind Gap, PA 18091; Ph: (610) 863-4161; Fax: (610) 863-8128.
New PA black slate, tools, hooks, nails.

*The above suppliers may also sell tools and accessories not listed. Call for complete information. New roofing slates, slate tools, and accessories may also be available through local roofing supply retail outlets. Used roofing slates may be available through local slate roofing contractors. Also refer to Chapter 11 for sources of tools and equipment. Check **index** for sources of finials, weather vanes, snow guards, and lightning rods.

Chapter Thirteen
Basic Repairs and Maintenance

There are a variety of routine repair and maintenance jobs that will keep a slate roof in good condition for generations if done properly. Our focus will be on repairs and maintenance above the *drip edge*, which is the very bottom edge of the slate where the water drips off during a rain. We will not include anything below the drip edge (such as rain spouting) in any detail, although rain spouting must be briefly discussed here because many spouting contractors nail rain gutters on top of slate roofs, which damages them. Flashing and chimneys will be covered more extensively in Chapters 14 and 15.

In order to best understand how to do routine repair jobs on slate roofs, one should first become familiar with the *parts* of the roof, which are the drip edge, the sheathing underneath the slate, the slate itself, the valleys, ridges, flashings and chimneys (see Figure 13.1). Let's start with the drip edge.

DRIP EDGE

The drip edge of a slate roof often needs repaired because contractors looking for a quick and cheap way to fasten gutters to a house simply nail them through the slate using strap hangers. This damages the slates along the drip edge, which eventually have to be repaired or replaced.

It's important to *not* allow strap hangers to be used on slate roofs *unless* the hangers are fastened *underneath* the slate. Originally, many slate roofs did use strap hangers nailed to the sheathing under the slate because the fascia boards on the older houses were not plumb (straight up and down), making it difficult if not impossible to fasten the gutters to the fascia. When the old gutters wore out and the old strap hangers rusted away, the contractors' solution was to nail

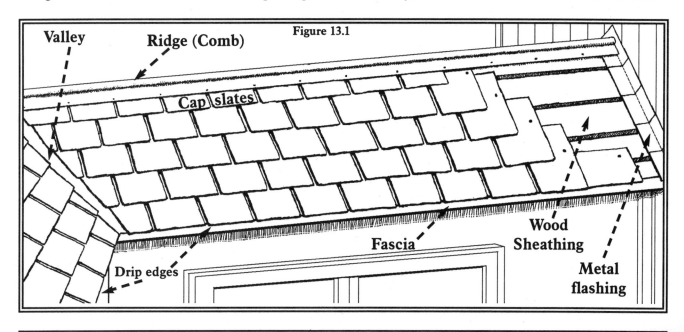

Figure 13.1

Valley

Ridge (Comb)

Cap slates

Fascia

Wood Sheathing

Drip edges

Metal flashing

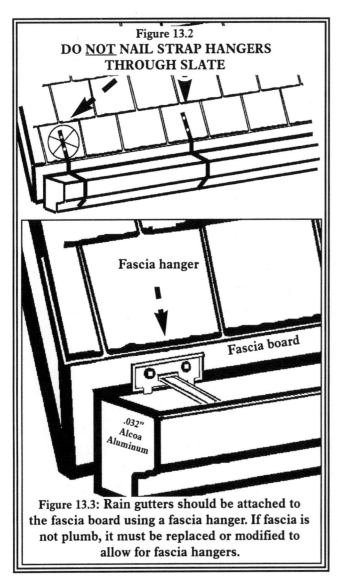

Figure 13.2
DO <u>NOT</u> NAIL STRAP HANGERS THROUGH SLATE

Fascia hanger

Fascia board

.032" Alcoa Aluminum

Figure 13.3: Rain gutters should be attached to the fascia board using a fascia hanger. If fascia is not plumb, it must be replaced or modified to allow for fascia hangers.

new strap hangers on top of the roof right through the slate, thereby damaging the roof.

The proper solution, however, is to rebuild the fascia so that it is made plumb, and then to hang the gutters on the fascia using fascia hangers (Figures 13.2 and 13.3). An alternative is to remove the slate where the hangers are to be installed, nail the hangers to the wood sheathing, then replace the slate over the hangers.

Often when repairing the slates along the drip edge of a roof, the gutter hangers must be removed one at a time in order to replace the broken slate underneath them. When the broken slate is off, the hanger can be nailed back directly onto the wood sheathing underneath the slate and the replacement slate then installed *over* the gutter hanger. The slate along the drip edge can be repaired this way without the need to take down the rain gutter beforehand.

A larger problem resulting from the unfortunate practice of nailing strap hangers through slate (the people who do this never have to repair the roofs; that's left to the slaters, and the roof owners must pay!) is the damage to the wood sheathing underneath the slate caused by the leaking roof. Because more water runs over the drip edge than any other part of the roof, this is the worst place for a leak to occur, and the wood underneath will eventually rot. In some cases not only do the slates have to be removed and replaced along the drip edge, but so do the boards along the drip edge. In severe cases even the rafter ends will be rotted and must be rebuilt. This is why roof owners must be vigilant in preventing unscrupulous contractors from ruining the drip edge of their roof by nailing strap hangers through their slate.

Much of the older spouting was the galvanized "half-round" type, hung on cast iron strap hangers fastened directly to the roof boards before the slate was laid. Today, however, aluminum spouting seems to be preferable to galvanized as it is much more durable, doesn't rust, and doen't need painted, although one must be careful to use the .032" thick aluminum (Alcoa makes a good gutter and hanger system) and not the thinner stock, which is not worth putting up. Die-hard restoration buffs will insist upon using the galvanized half-round spouting when replacing old gutters, as the author has done. After watching new galvanized spouting rust in only five years, the author has reluctantly come to the conclusion that heavy gauge aluminum spouting is definitely superior to galvanized. Perhaps the older galvanized spouting was made better than today's galvanized. Or, in other words, they don't make it like they used to!

An alternative to aluminum spouting is copper spouting, which is relatively expensive but quite durable, or stainless steel, which is hard to get *and* expensive, but exceptionally strong and

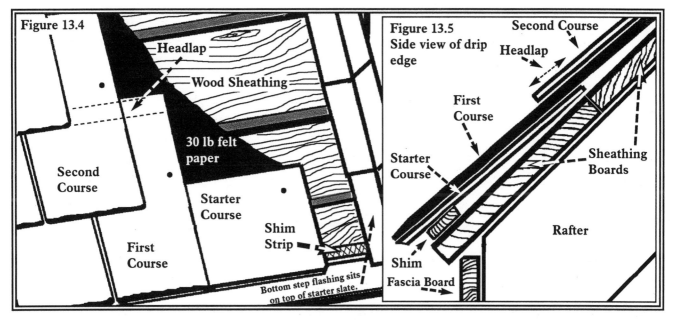

Figure 13.4

Headlap

Wood Sheathing

30 lb felt paper

Second Course

Starter Course

First Course

Shim Strip

Bottom step flashing sits on top of starter slate.

Figure 13.5
Side view of drip edge

Second Course

Headlap

First Course

Starter Course

Sheathing Boards

Rafter

Shim

Fascia Board

durable. In England, heavy cast iron spouting is popular.

OK, so your drip edge is bad and you have to repair it. It's not hard if you know how a drip edge is put together. The bottom course of slate is called the *starter course*, and it is usually laid sideways, as in Figure 13.4, above. It is an invisible row of slate as it lies *under* the first course. Under the starter course is a *shim*, which is usually a narrow strip of wood approximately 1/2" to 5/8" thick, nailed to the sheathing and running horizontally along the bottom edge of the roof. The purpose of the shim is to cock the starter slate at an angle so it matches the angle of all the rest of the slate on the roof (Figure 13.5). A good shim would be made of cedar or other rot-resistant wood, although almost any solid wood strip will work, and it's convenient to make the shim out of the same wood as the sheathing is made from.

After the starter course, the remaining slates are nailed in the standard way allowing for at least a 2" headlap, up to 4" on lower sloped (4:12 or 4" of rise in 12" of run) roofs. *The headlap is the amount of overlap each slate has in relation to the <u>second</u> course of slate above (or below) it.* Every slate is overlapped by the course above it, but it's the overlap on the *second* course that's critical (Roof slope and slate installation are discussed in greater detail in Chapter 12).

As with any repairs, repairing drip edges requires removal and replacement of the bad slates (which will be cracked, broken, or tarred), and removal and replacement of rotten boards. The wood sheathing traditionally used on slate roofs is one inch thick, rough sawn (unplaned), solid lumber usually from a local source, and usually installed "green" (not kiln dried). Local, green lumber can't be bought at standard lumber yards, but is available at sawmills, which are abundant in any forested area (such as in the northeast US). Plywoods, laminated woods, and particle boards are <u>not</u> recommended for slate roofs as these materials can (and do) de-laminate (come apart) over time. Local, green lumber is easy to get, costs less, and is superior to laminated wood in strength and durability, so use it if you want to do a long-lasting job. Green lumber does <u>not</u> need to be dried before using. This issue is discussed in greater detail in Chapter 12.

It should be added that many older slate roofs have or had built-in gutter systems on the roofs. Many of the old built-in gutters have been removed because they weren't properly maintained (the metal wasn't kept painted), and they rusted away and leaked. You will see many old slate roofs with a couple of layers of asphalt shingles along the bottom edge, because the people who removed the built-in gutters covered the exposed sheathing with asphalt shingles instead of slate as they were supposed to. These roofs can be restored by removing the shingles and reslat-

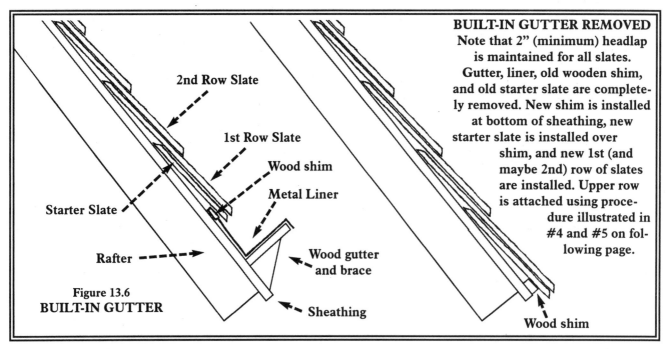

BUILT-IN GUTTER REMOVED
Note that 2" (minimum) headlap is maintained for all slates. Gutter, liner, old wooden shim, and old starter slate are completely removed. New shim is installed at bottom of sheathing, new starter slate is installed over shim, and new 1st (and maybe 2nd) row of slates are installed. Upper row is attached using procedure illustrated in #4 and #5 on following page.

2nd Row Slate

1st Row Slate

Wood shim

Metal Liner

Starter Slate

Rafter

Wood gutter and brace

Figure 13.6
BUILT-IN GUTTER

Sheathing

Wood shim

ing the drip edge (as above).

THE BASIC REPAIR

The basic repair job on slate roofs involves the removal and replacement of individual slate. Slates will be removed because they're broken or tarred, but they'll also be removed when they're covering flashing or sheathing that must be replaced. In almost any slate roof repair situation, slates must be taken off the roof and then put back on.

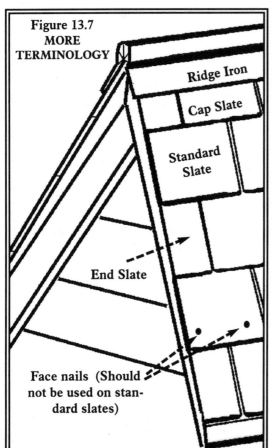

Figure 13.7
MORE TERMINOLOGY

Ridge Iron

Cap Slate

Standard Slate

End Slate

Face nails (Should not be used on standard slates)

Roof slates can fall into six general categories, depending on their position on the roof. At the bottom of the roof are the *starter* slates, which we have already had a look at (Figure 13.4). Above the starter slates are the *standard* slates, which are full size, uncut slates. At the top of the roof under the ridge iron are the *cap* slates, which are relatively small. At the gable ends of the roof are *end* slates or half slates, which are cut approximately in half lengthwise (see Figure 13.7).

Then there are the *valley* slates, or *flashing* slates, which are cut to any size or shape to fit against or on top of flashing, and finally we have *ridge* and *hip* slate, which are used as ridges and hips (also known as *saddle* or *comb* ridges or hips) in place of metal ridge.

The basic slate repair involves the standard slates, which are the full size, uncut slate that make up the bulk of the roof. If one of these slates breaks off, it must be removed and replaced. This is simply done by pulling the two nails out of the roof which are holding the slate in place by using a slate ripper, then sliding a

Basic Slate Repair

Figure 13.8

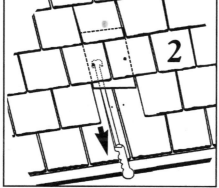

A common sight on old slate roofs - a broken slate. The roof will leak here at the two exposed nail heads, and in the space between the lower slates, near the top of the slot.

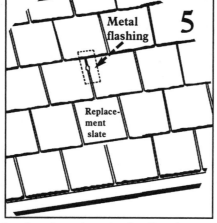

#2 shows the hidden part of the broken slate underneath the overlying slates, and the location of the nails holding it. The nails are pulled out from under the slate with a ripper and a hammer (see page 171).

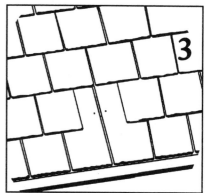

#3 shows the broken slate and nails removed and the space ready for a replacement slate, which will be slid in, top up and front facing out (Figure 12.8, page 200)

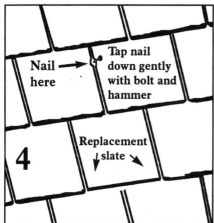

Nail here → Tap nail down gently with bolt and hammer

Replacement slate

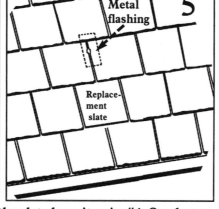

Metal flashing

Replacement slate

The new slate is nailed through the slot above it as in #4. One hot-dipped galvanized or copper nail 1+ 1/2" long will usually suffice. Chip a hole in the slot first to allow room for the nail head. The nail must be "set" by tapping with a common bolt so the nail head is below the overlying slates. Then the nail head is covered with painted aluminum or copper flashing as in #5. Slightly bend flashing lengthwise before inserting. Refer to figures 11.14 and 11.15, page 183.

DON'T DO THIS

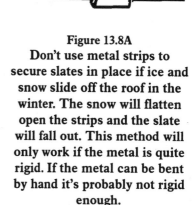

Figure 13.8A
Don't use metal strips to secure slates in place if ice and snow slide off the roof in the winter. The snow will flatten open the strips and the slate will fall out. This method will only work if the metal is quite rigid. If the metal can be bent by hand it's probably not rigid enough.

Figure 13.8B
USING A SLATE HOOK

Old slate is removed and slate hook is nailed in place

New slate is held in place by slate hook

See pages 183-184

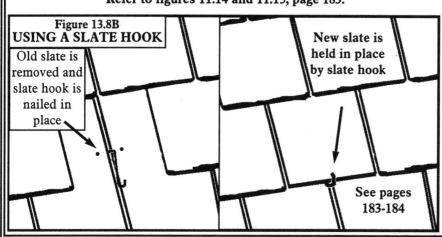

new slate in place and nailing it with one 1 1/2" hot-dipped galvanized (or copper) nail, through the slot overlying the slate. Sometimes two nails must be used to get a solid replacement job (when replacing a row of slates, every few slates should be double nailed to prevent the slates from becoming crooked over time). The nail head is then covered with a piece of metal flashing (called a *bib* flashing) slid under the slot (but over the nail) to make the repair leak-proof. You may wish to refer to Figures 11.2 (page 171), and 11.14 - 11.15 (page 183), which illustrate this process, in addition to Figure 13.8 on the previous page.

A common alternative to this method involves the use of the *slate hook* to hold the replacement slate in place (see Figure 13.8B)

It's common for some roofers to just "face-nail" replacement slates, which means they drive a nail or two through the face of the slate and leave them there, exposed to the weather. Sometimes the nail is gasketed, sometimes it is simply covered with roof cement or caulk, and sometimes it is neither gasketed nor covered. Usually, this is <u>not</u> the right way to do the job because face nails, including gasketed ones, will eventually leak. An exception is made in the case of dilapidated *soft-slate* roofs, when the roof is scheduled for replacement and the caulked face nails may temporarily help hold the roof together.

Contractors face-nail slates because it's easier and cheaper than hiding the nail in a slot and flashing over it, or using slate hooks, and most homeowners don't know whether the contractor's doing a good job or not, so they're taken advantage of.

Some slates, such as cap slates and end slates, don't have slots to nail through, and end slates are the only acceptable slates for face nailing on hard slate roofs, and even they don't *have* to be face-nailed (Figure 13.9). Let's take a look at the various slates and how to fasten them in place when replacing them:

In order to replace a *starter* slate, the overlapping slates usually need not be removed. The starter slate can be pulled out with the ripper by hooking its nails and pulling them out, thereby removing the old slate. Then a new starter can be inserted underneath the first row of slate and nailed in place through an exposed slot (or two). Sometimes the overlying slates must be removed in order to get a starter slate out. Once the overlying slates are removed, the starter slate can easily be nailed in place,

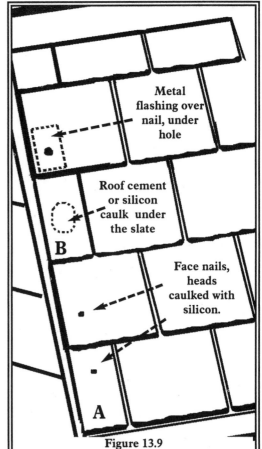

Figure 13.9
Two Ways to Replace End Slates
Slate "A" was fastened using two 1& 1/2" long hot-dipped galvanized or copper roofing nails. The nail heads were caulked with clear 50-year silicon caulk. Slate "B" was bedded in a little roof cement (underneath the slate), and nailed through a hole punched in the overlying slate with the point of a slater's hammer. The nail was covered with a piece of metal slid under the slate with the hole, but over the nail. The metal can be bedded in a bit of cement too, to keep it from falling out.

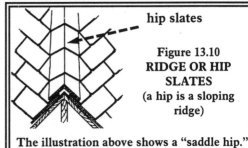

hip slates

Figure 13.10
RIDGE OR HIP SLATES
(a hip is a sloping ridge)

The illustration above shows a "saddle hip." These slates must be removed when replacing the underlying slates.
[From <u>Slate Roofs</u>, 1926, p. 19]

then the overlying slates replaced, with *their* nails hidden in the slots. Starter slates lay on a shim strip, and that shim may need to be replaced (at least in part) in order for the starter slate to lay properly.

Cap slates do not have exposed slots allowing for replacement nailing in the standard fashion, so they must be nailed under the ridge iron by prying the ridge iron loose and lifting it enough to get a nail under it. The ridge iron has to be pried loose anyway to get the old cap slates out. In some cases the cap slates can be fastened by nailing right through the ridge iron with an 8 penny nail, then caulking the nail head with roof cement or caulk (50 year clear silicon caulk is recommended). The disadvantage to this technique is that when the ridge iron is removed for replacement, those cap slates will fall out. When the ridge or hip is made of slate, the slate must be removed in order to replace underlying slate (Figure 13.10).

End slates are only half as wide as the standard slates and don't have a slot over them to nail through. Here is where a face-nail is acceptable. Use the standard 1 1/2" hot dipped galvanized (or copper) nail, poke the nail hole in the slate with a slate hammer before nailing so you don't break the slate, then caulk the exposed head with 50 year silicon after the repair has been made. Or, bed the end slate in roof cement before face-nailing it by placing a dab of cement *underneath* the slate where it can't be seen, and where it will keep the face-nail from leaking. An alternative to face-nailing an end slate is to chip a hole in the overlying slate, nail through the hole, and flash under the hole and over the nail head (see Figure 13.9).

Gable-end slates have the unfortunate reputation of easily blowing off some roofs (such as old, wind-exposed barns). When gable-end slates are being replaced on a roof that suffers from chronic wind damage, the slates should be bedded in roof cement. This is done by spreading roof cement <u>under</u> the slate before the slate is nailed in place, so the slates are glued together like a series of peanut butter sandwiches, and no roof cement is visible. This little trick will keep those gable-end slates from blowing off again (Figure 13.11).

Most *flashing* slates, such as slates along a valley, have a slot overlapping them, allowing a place for a standard replacement nail. Otherwise, the overlying slates must be removed, the flashing slate replaced, then the overlying slates replaced in the standard fashion (i. e. nailing through the overlying slot, then sliding some metal in). When small slates are replaced it's often a good idea to bed them in a little roof cement or silicon before nailing. Poke a hole in the small slate with a slate hammer, cutter punch, or other sharp object, put a dab of roof cement *under* the slate where the hole is, position the slate on the roof, then nail in place. The cement helps hold the slate in its proper position, and prevents any leaking around the nail.

Figure 13.11
WIND-EXPOSED EDGES ON OLD BUILDINGS

Some buildings, particularly barns, are exposed to the wind, and the windward side repeatedly loses its slate along the gable end during heavy winds, leaving a ragged and damaged roof edge. When these slates are replaced, they should be bedded in trowel grade roof cement, so that the cement is sandwiched between the slates and is not visible. This simple technique tightens up the windward edge and prevents the slates from rattling loose again.

TEMPORARILY SEALING LEAKING VALLEYS

There is nothing more aggravating than a roof that leaks. Probably the most common source of leaks on slate roofs is broken or missing slates as well as slates with holes and/or hidden leaks (which we will discuss shortly). Valleys are another common source of leaks. Valleys carry more water than any other part of the roof, as they act as a channel collecting water from two roof planes. When the valley wears out, holes develop which can leak a large amount of water into a building, and a pinhole in a valley can leak buckets of water. The solution to leaking valleys is simply to remove the old valley metal and replace it with new metal. This is a routine job, discussed in Chapter 14, that is a sure and permanent cure for any leaky valley, especially if the replacement metal is copper, stainless steel, or heavy gauge aluminum.

Sometimes the roof is not worth the cost of replacing entire valleys because the slate is soft (see Chapter 6) and nearing the end of its life. In this case, simply spread trowel-grade roof cement on the valley metal, roll out some fiberglass roofing mesh onto the cement, work the mesh in, then coat it with another layer of roof cement, being careful to make sure *no cement overlaps the slate*. Repeat, the cement should be worked *under* the slate with a trowel, <u>not</u> over it (Figure 13.12).

There are thousands of roofs with leaking valleys where a homeowner or roofer has tarred both the valley *and* the slate on both sides of the valley. This defaces the slate and ruins the appearance of the roof, while adding nothing to the effectiveness of the valley repair. If anything, it will make the valley leak worse over the long run, because the water will run down the roof and seep under the roof cement, which will act like a dam drawing the water into the roof. It is imperative to work the cement *under* the slate, then when the valley is eventually removed and replaced, the slate will still be good, and the valley replacement job will be much easier.

This three-step cement and fiberglass repair method is an inexpensive temporary measure to be used when the roof is not expected to last much longer, or in situations where the roof owner doesn't have the money to pay for a permanent roof repair. A roof cement and fiberglass repair may last 10 years, and can also be used to repair almost any leaking surface (flat roofs, metal roofs, chimney flashings, and built in gutters, for example) *except* slate surfaces. Eventually, the roof cement and fiberglass repair will wear thin, but it can easily be redone and

Figure 13.12
TEMPORARY BUT EFFECTIVE WAY TO SEAL A LEAKING VALLEY
(and other non-slate leaking surfaces)

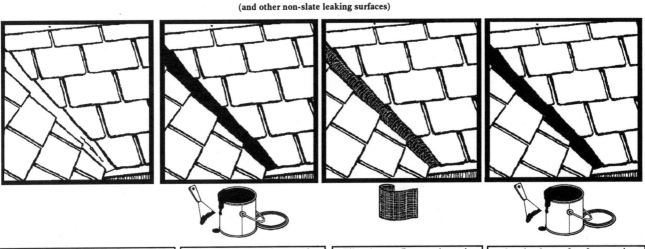

The valley metal is cracked and leaking, but the slate is soft and it's not worth replacing the valley, or the roof owner can't afford to replace it.	So trowel grade roof cement is applied over the entire length of valley metal and worked <u>under</u> the edges of the slate.	Fiberglass roofing membrane is rolled into valley, worked <u>under</u> the slate edges and troweled into the cement	Another layer of roof cement is applied over the fiberglass carefully, so as not to get any on the slate.

will last quite some time this way. Valley *replacement* is discussed in detail in the next chapter, so we won't dwell on it any longer here.

REPAIRING HOLES AND "HIDDEN LEAKS"

Don't Use Tar

When valleys leak, the solution is simple - replace them as described in the next chapter. If you can't afford to replace them, seal them. Yet there are often leaks on slate roofs that aren't in the obvious places (valleys, chimneys, flashings, and missing slates) and are hard to pinpoint, and this is where a bit of experience comes in handy. Being able to find the source of leaks (and repair them, of course) is the true mark of a professional, because many leaks are caused by something small and can be repaired rather easily.

Many, many roofs have tar spread all over them because the repair person didn't know how to look for and locate the source of the leak. Instead, he got out the tar bucket and the brush and went wild, hoping he'd hit the leak, which was probably the size of a pinhole. This is somewhat equivalent to spreading tar all over your car because there's a slow leak in a tire.

Quite frankly, there is no greater folly than painting a slate roof with tar, yet, unbelievably, some contractors actually *advertise* the "service" of coating slate roofs with brush-on roof tar! Looking at the bright side of this, contractors who advertise that they coat slate roofs with tar are also advertising their incompetence, so you know automatically not to hire them for anything. On the down side of this, however, many homeowners are talked into having their slate roof painted with tar because the contractor was a smooth talker, and it "sounded good" to the homeowner. The tar doesn't ruin the roof functionally, but it does ruin it aesthetically, and it takes about 50 years to wear off. In the meantime, besides having to look at an ugly roof for the rest of your life, if the roof does need repaired (and it will - that's why it was coated; it leaked!) the job of repairing it properly is much more difficult when the roof is all glued together with tar. If you run into a contractor offering to tar your slate roof, smile, speak gently, humor him, and get rid of him as soon as possible. Better yet, photocopy this page and give it to him as a present (and suggest he get a copy of this book for himself).

Holes in Slate

One common source of a leak on a slate roof is a hole in a slate. Holes often result from nails that weren't nailed down far enough when the roof was installed, or which backed out of the roof as the sheathing boards dried, and the nail heads worked against the overlying slate eventually to poke a hole right through (see Figure 13.13). If you examine a roof closely, you'll likely find these holes, and the guilty nail heads will be happily peeking through them. They're easy to fix. Simply take your hammer and nail set (a 1/2" by 6" common threaded bolt does fine as a nail set) and tap the nail head down where it's supposed to be, then slide a piece of metal (copper or painted aluminum - brown side facing out) under the hole and over the nail. A good size for the metal is 4" wide and 7" long, and it should have a slight bend in it lengthwise so it doesn't slide back out. You can fix a lot of holes on a slate roof quickly and inexpensively this way. On very old roofs there may be fifty or more holes like this. Alternatively, the slate with the hole can be completely removed and replaced.

Holes in slate can be caused by other things too. One farmer thought it was a good idea to shoot the pigeons that roosted in the rafters of his barn. His .22 rifle might not have killed many

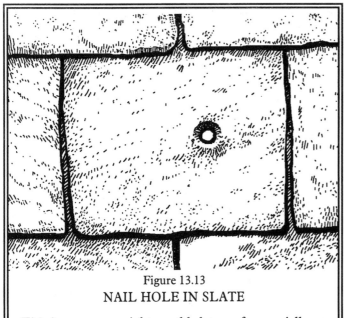

Figure 13.13
NAIL HOLE IN SLATE

This is a common sight on old slate roofs, especially on soft slate roofs from Pennsylvania's Lehigh-Northampton district as above. To repair, simply tap the nail down a bit with a bolt, and slide a piece of copper or aluminum flashing under the slate and over the nail.

pigeons, but it sure made a lot of neat little holes in his slate roof.

Then there will always be holes in slate roofs that have no obvious explanation. Maybe someone fired a gun in the air a mile away and the bullet hit your roof, or maybe it was a miniature meteorite, or maybe a kid threw his dad's screwdriver at a bird on your gutter, or maybe . . . Well, who knows why, but there are often holes in slate roofs that are inexplicable. Again, a piece of flashing slid <u>under</u> the slate will solve the problem quickly and easily. Do not tar over the hole unless you really like looking at ugly things.

Hidden Leaks

This is where the fun begins. Hidden leaks are leaks that roofing contractors "fix' and they still leak. Many roofing contractors aren't very well versed in the art of finding hidden leaks, and after two or three attempts they just give up, leaving the roof owner pulling his or her hair out and fuming. Ironically, most hidden leaks are *caused by* roofing contractors. That's probably why they can't find them.

Numerous examples spring to mind. One homeowner who was building his own house subcontracted the roofing out to a roofing contractor, who also flashed the chimney that protruded through the roof. Well, after the flashing job, the roof leaked around the chimney. The owner called the contractor back and he "fixed" it. It still leaked. He called him back again, he "fixed" it again, and, of course, it still leaked. At this point the author was called in to have a look at the problem (gratis), and the leak was immediately obvious - the chimney wasn't flashed properly. No amount of jerry-rigging was going to make the chimney flashing waterproof if it wasn't installed properly to begin with. The only solution was to tear it out and do it again. In this case, the roofing contractor didn't know the secret to flashing a chimney (folding the corners, which is explained in Chapter 14 of this book). In the end, the contractor agreed to let the homeowner get someone else to reflash the chimney and he promised to pay the bill, neither of which had yet been done a full year and a half after the original flashing job.

Bad flashing jobs on chimneys are all too common, but a more widespread cause of hidden leaks on slate roofs is *faulty old repairs*. If you've been reading this book and paying attention, you know by now that when you nail a replacement slate in place, you hide the nail in the overlying slot and cover it with non-corrosive flashing. Unfortunately, many roof repair-people don't do that. Instead they create problems:

1) Problem: they face-nail slates. These are nails just nailed through the exposed face of the slate and left exposed to the weather. They'll eventually leak around the nails. Sometimes the face-nails are gasketed, sometimes they're cemented over, sometimes they're caulked. Eventually they'll all leak. Face-nailed slates are obvious, however, and therefore easy to find and remove, although a lot of people don't realize that they're the cause of leaks. The heads of very old face-

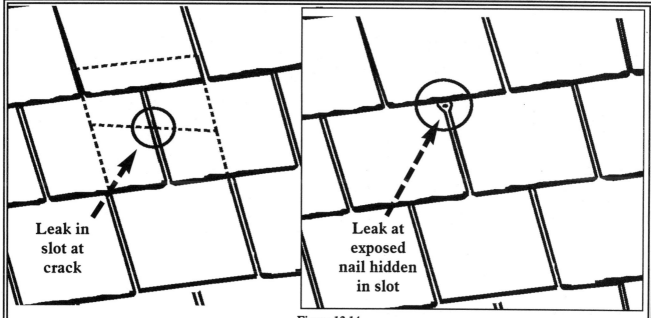

Leak in slot at crack

Leak at exposed nail hidden in slot

Figure 13.14
Two kinds of hidden leaks. The one on the left shows a slate that is cracked underneath the overlying slate. It will leak where the crack crosses the slot. The one on the right shows an old nail head hidden near the top of the slot. These can be hard to see, and they will leak.

nails will completely rust away and the nails will be very hard to find unless the roof is examined closely. On the other hand, some *soft-slate* roofs are so rapidly falling apart that face-nails are the only thing holding the roof together. When an old soft-slate roof only has five or ten years left, face-nailing may actually help it, provided the nail heads are kept caulked or gasketed.

Solution: On hard slate roofs remove the face-nails and the slate that was face-nailed and replace the slate. This time nail the new slate in properly. On *soft slate roofs* (especially PA ribbon slate), caulk the nail heads with 50-year clear silicon caulk and don't try to take the slates out. Chances are that if you start taking a slate or two out, the entire roof will start falling apart and you'll have created a major headache for yourself.

2) Problem: they use corrosive flashing. When nail heads are covered by flashing that will rust, eventually it does rust away and the exposed nail head, even though tucked down in the slot, will leak.

Solution: Use your ripper to remove the corroded flashing and replace it with copper or painted aluminum or other non-corrodible flashing material. When using painted aluminum, use a brown color and leave the brown facing out.

3) Problem: they don't cover the nails. It's not uncommon for a roofer to replace a slate, nail it in the overlying slot like he should, then leave it like that without covering the nail with flashing, as if his brain was working fine and then just shut down before the job could be finished. This is one of the hardest hidden leaks to find, because the nail is tucked in right at the top of the slot where it's hardest to see (especially from above). Then the nail rusts until it matches the color of the roof, and becomes invisible.

Solution: When examining the leaking area, always look in the slots for hidden nails. Look closely. If you find one, renail the slate and flash over the nail.

4) Problem: they cover the nail heads with roof cement. Some roofers nail the replacement slate in the slot, then put a dab of roof cement over the nail head instead of flashing. The roof cement then wears away and the nail leaks. These leaks can be hard to find because there isn't

much to see, and like the previous example (#3), close examination of the slots in the leaking area will flush the culprit out. *Solution: renail and flash* (Figure 13.8).

As long as we're on the topic of leaks hidden in the slot, we may as well mention that sometimes slates crack widthwise; the crack is up underneath the overlying slate where it can't be seen, and a leak develops where the crack crosses the slot. This kind of leak can only be found by closely examining the slots, while maybe wiggling the slate a bit. You must get your face right down to the slot to see a crack inside it (Figure 13.14).

Capillary Action

When liquids are in contact with solids, an adhesive force exists between them that may cause the liquid to creep in one direction or another along the surface of the solid, independent of the force of gravity. Water will run sideways or even uphill. This is an extremely important principle to be aware of when trying to understand leaks on roofs, and it pertains particularly to leaks around flashing. The rule of thumb is that water has the capacity to run sideways 2", which is why the bib flashing (the flashing that covers the nail on a replacement slate) is

Figure 13.15
NEVER TAR THE SLOTS ON A SLATE ROOF
as shown above on a historical house in Mercer, PA, with excellent sea green slate. This is a popular way to ruin a roof aesthetically, and the tar does not repair the roof. If you think there may be a leak in a slot, *slide a piece of metal under the slot, do not tar over it!* In the photo above, someone in the past had walked on the slates, evidently cracking one, and it subsequently leaked in the slot. The repairperson's "solution" - tar <u>all</u> the slots!
(Photo by Barry Smith)

4" wide. If you have a situation on your roof where water can run sideways and find a place to leak in within that 2" limit, you may have a leak by capillary action. If you do have a leak by capillary action, your roof will probably look like there is absolutely nothing wrong with it, and you will be driven crazy trying to find the source of the leak. The solution is to examine the roof closely where it's leaking and look for a place where the lateral overlap on the roof is less than 2". This may be a slate overlapping a piece of flashing, or a slate overlapping another slate. If you find such a situation, *slide more flashing metal under the guilty area to reinforce it*. That should easily solve the problem.

Water Travels (Usually) Downhill

A house with a T-shaped slate roof had a chimney near the center of the roof. A leak developed in a bedroom ceiling, but the slate roof above the bedroom showed no sign of a leak. A close examination in the attic revealed that the water was leaking in around the chimney, running down a hip rafter, detouring down a jack rafter toward the bedroom, dripping on the ceiling's edge, then pooling in the center of the ceiling (because the old ceiling joists were slightly sagging). The water then dripped annoyingly into the room. Repair of the chimney flashing, located on the other side of the house, stopped the leak. If a leak shows up in your roof, you must understand that the water can be entering the roof from any point above the leak.

I've Tried Everything and I Still Can't Find the Leak

Don't despair. The author boasts that he can find any leak on a slate roof, and every now and then he runs into a tough one. The solution is not convenient but it's simple: wait till it's raining and get up into the attic of the house. Look for the leak, trace it back to its origin, mark the spot with a piece of chalk or a thumbtack and rag, or anything prominent, then call your roofing contractor. He'll come in dry weather, climb into your attic, find the source of the leak, go up on the roof and fix it. This works.

I was called to a house where the chimney suffered from chronic leaking around the flashing. I asked the homeowner if the chimney was used for anything, and when he replied that it was not, I suggested the chimney be removed to below the roof line (a routine job). He agreed, and I removed the chimney, closed up the hole with matching lumber (1" rough sawn local lumber), and slated the area to match the rest of the roof. When the job was done, you couldn't tell by looking at the outside of the roof that there had ever been a chimney there. I assured the homeowner that, with the chimney gone, the chronic leak had been brought to an end.

A few weeks later the homeowner called me and said it was still leaking. He said he could hear the water dripping in the roof when it rained. So I went back the next day, got up on the roof and looked at the area closely. I could see nothing that might be leaking, but I double checked everything and reinforced it with some extra flashing before I got down. "I don't see any leaks," I said, "and I've never had a chimney removal job leak before," I added, "but it should be OK now."

A couple weeks later he called me again, madder than a wet hornet. "It's still leaking! I heard it dripping in the roof last night in the rain!" He wanted me to drop everything and get over there right away; after all, he had already paid me to fix his leak and he was suggesting that I had ripped him off. I went back the next day, looked it over again, saw nothing that would be leaking, and finally told him, "You have a *hidden leak*. I can't see any sign of anything that would be causing a leak. The only way to fix this thing is to wait for it to rain again, get up *inside* your roof, go to where the chimney is (it was still standing in the attic), and mark the spot on the underside of the roof sheathing where it's wet. Call me when you've done this and I'll go up into your roof, find the spot you've marked, and *then* I'll know where this damn leak is and be able to fix it!"

Well, it wasn't more than a week before we had torrential rains. The rain just beat down in sheets for hours, and you could really hear it pounding on the roof. I knew that the homeowner would be calling, or so I thought. He didn't call, however, and more rain came, another week passed, and he still didn't call.

Then, a month later I ran into the guy at a grocery store. He looked at me sheepishly as if he hoped I wouldn't see him. I said hello to him; he returned my greeting, then took me aside and said, "You know that leak I had?" "Yes." "You know those real heavy rains we had a few weeks ago, when the rain just beat down like hell?" "Yes." "Well, I did what you said, I went up into the roof with a flashlight." "Yes?" "And it was just raining like *hell*, the sound of the rain beating on the roof was deafening." "Yeah? Did you find the leak?" "Well, you know what, I watched that spot on the roof above the chimney for an hour with my flashlight, and guess what - *it didn't leak a drop*. It was just as dry as a bone."

At this point unseemly thoughts came to my mind, but I remained polite and made a mental note to avoid this guy in the future. The job had been done right in the first place, but the man heard some dripping during a rainstorm and immediately concluded that it was dripping where the roof had been repaired. It was probably dripping outside, but nothing would convince the man until he got up into the roof during a rain and saw for himself.

The moral of the story is this: if you have a leak and can't for the life of you figure out where it's coming from, get into your attic, if possible, and look at the underside of your roof on a rainy day. Who knows, you may find that the leak is a figment of your imagination! Otherwise, you'll pinpoint the location of the leak and, if you've marked it well enough so it can be found again on a dry day, it can be readily repaired.

On another occasion a fellow called me to complain that he had a leak in the same place on his ceiling for fifteen years. He said he's had a whole slew of roofers try to fix it and no one ever could. Of course, nothing interests me more than a challenge like this, so I stopped by the guy's house and took a look at his roof. Sure enough, the area directly above the leak was sound. There was nothing that looked like it would be causing water to get through the roof. After fifteen years of roofers trying to find this one, I decided to put this leak directly into the "hidden leak" category, and find it from the inside of the house.

This was a bit of a problem because we had to remove a dropped ceiling and then cut through a wooden ceiling to get at the underside of the roof. I asked the homeowner to call me the next time he saw any leaking occurring, and soon thereafter, on a rainy day, he called. I drove through the rain to his house, climbed up into the ceiling with a flashlight, and traced the path of the water up the roof about ten feet and sideways about four feet to the point of origin. I measured the distances, wrote them down, and later, during dry weather, I climbed on the roof and went to the leaking spot like finding treasure on a treasure map. The water was coming in at a hip because the roof slates were cut too short against the hip when the roof was installed, and the hip slates covering them didn't have enough overlap (remember capillary action). All of this was impossible to see until the roof was torn apart. The water traveled from the hip down a rafter, then ran sideways on a sheathing board, crossed two rafters and finally dripped through the ceiling far from the point of origin. This ornery little leak confounded many a roofer and drove the homeowner crazy for fifteen years, but was easily repaired once the mystery had been solved.

CHIMNEYS

Chimneys are a common source of leaks on a slate roof. They leak around the flashing, which is the metal that seals between the roof and the chimney itself. The flashing eventually rusts, develops holes, then leaks. The solution is to replace the flashing, which is a routine job on old slate roofs, a job that should be done once every 75 years or so. A good material to use when replacing chimney flashing is 16 ounce copper. We will not go into the reflashing of chimneys in any depth here, because we focus on that topic in Chapter 14 (and on brick laying in Chapter 15).

However, there are people who don't want, or can't afford to pay to have their chimney reflashed, but still want to stop it from leaking. Although reflashing is the only permanent cure, the chimney can be sealed at the flashing with roof cement and fiberglass membrane as a temporary measure. Such a seal will last about ten years, then will have to be redone, or the chimney will have to be reflashed at that time.

Sealing a chimney at the flashing is done in the same manner as sealing a valley. The three step process of applying trowel grade (black, plastic) roof cement over the flashing, working in a layer of fiberglass mesh, then applying another layer of roof cement, will create a water-

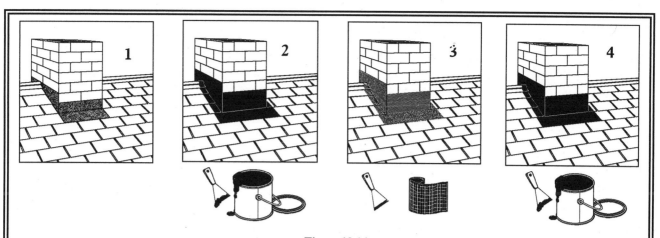

Figure 13.16

TEMPORARY SEAL AROUND CHIMNEY FLASHING.

It's better to reflash a chimney than it is to tar it up, but for those who don't have the means, this will do in a pinch. See Chapter 15 for information on how to reflash a chimney.

1. Original chimney and leaking flashing.
2. Loose flashing is removed. Flashing area is thoroughly, but neatly, covered with trowel grade roof cement, which covers all flashing and overlaps roof slates on sides of chimney about three inches .
3. The roof cement is then covered with a layer of black fiberglass roof membrane available at better hardware stores and building supply outlets in 6" or 12" rolls (membrane is trimmed with a razor utility knife). The membrane is folded 90 degrees to cover joint between roof and chimney in one piece.
4. The fiberglass is then covered with a thin 2nd layer of roof cement.

proof seal that will last for quite some time. The problem with this process, however, is that it makes the reflashing job considerably more difficult when that job is finally gotten around to, because all that roof cement must be removed. And so it is with some hesitancy that I even recommend sealing chimney flashing with roof cement and fiberglass. On the other hand, such a seal is quick and cheap, so a lot of people do it anyway, and if you're going to do it, you might as well do it right.

A roof cement and fiberglass seal doesn't need to be ugly; it can be neatly done, and the cement can even be painted with tinner's red to make it look like real flashing from a distance. When applying the cement, use a three inch square edged trowel preferably with a flexible blade. Apply the cement over the flashing on the sides of the chimney *and* on the slate roof a distance of 3" from the chimney. Use at least a 6" wide roll of roofing fiberglass, placing the fiberglass so it lays on the roof 3" and runs up the side of the chimney another 3" (minimum). It's critical that the fiberglass be folded at a 90 degree angle to lay on the roof and on the side of the chimney at the same time in a single piece, because the area most likely to leak is the crack where the chimney and roof join (Figure 13.16). Roof cement will dry and crack over the years and water will leak right through it if you don't reinforce the joints with the fiberglass. You want to run both the cement and the fiberglass up the side of the chimney far enough to cover the existing flashing, so you may have to slightly overlap the fiberglass on the chimney sides if applying a second layer. If the flashing on the side of the chimney is loose, remove it before sealing the chimney.

Use the roof cement liberally, but don't spread it anywhere it isn't necessary. It's hard to remove from bricks, and excess roof cement will look ugly when the chimney is finally reflashed. Be aware that when you put roof cement on top of slate, those slates are aesthetically ruined and should be replaced when the chimney is finally reflashed.

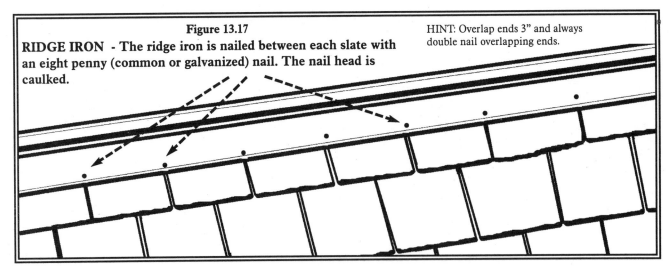

Figure 13.17

RIDGE IRON - The ridge iron is nailed between each slate with an eight penny (common or galvanized) nail. The nail head is caulked.

HINT: Overlap ends 3" and always double nail overlapping ends.

LEAKS AT RIDGES

There are several types of ridges popularly used on slate roofs, but these can be lumped into two basic categories: metal and slate. (The ridge is the horizontal peak of the roof, also called the "comb".)

Metal ridges are far more common than slate ridges, and are typically made of galvanized steel, although both copper and aluminum ridges are also available. The common galvanized ridge metal, also known as "ridge iron" by slaters, and "ridge roll" by manufacturers, will rust and deteriorate if not kept painted. The quality of ridge iron has progressively dropped over the years, and it's hard to find anything worth putting on a slate roof, although Benjamin-Obdyke Co. in Warminster, PA, makes a good, 26 gauge, galvanized ridge roll for slate roofs. Ridge iron should be 26 gauge galvanized (lighter, flimsier gauges are 28 and 30 gauge, which should be avoided if possible - especially the 30 gauge), and it must be painted approximately every five years with tinner's red oxide, the traditional roofing paint. A well made ridge iron is "V" shaped to conform to the shape of the peak of the roof.

When ridge iron is neglected, it rusts and will eventually leak. When that happens, the solution is to replace the ridge iron with new, 26 gauge, galvanized ridge (watch out for wasps, bees and bats under ridge iron). Traditionally, the ridge iron is nailed in place with 8 penny common or galvanized nails, and the nail heads are cemented with roof cement or caulked with 50-year silicon. The ridge iron is nailed in every slot between the slates it's overlapping (see Figure 13.17). Don't use screws on ridge iron because old rusty screws are very difficult to get out without damaging slates when it's time to replace the ridge metal.

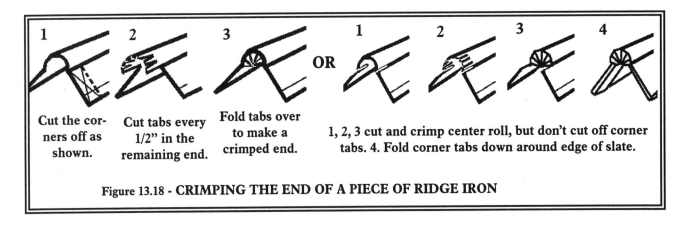

1. Cut the corners off as shown.
2. Cut tabs every 1/2" in the remaining end.
3. Fold tabs over to make a crimped end.

OR

1, 2, 3 cut and crimp center roll, but don't cut off corner tabs. 4. Fold corner tabs down around edge of slate.

Figure 13.18 - CRIMPING THE END OF A PIECE OF RIDGE IRON

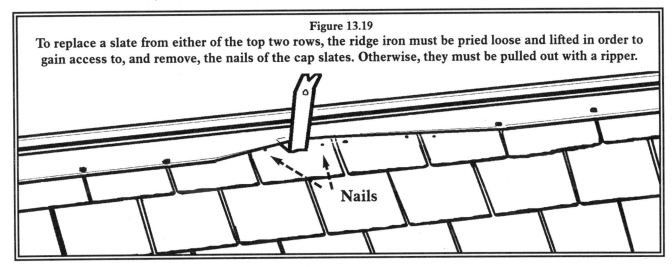

Figure 13.19
To replace a slate from either of the top two rows, the ridge iron must be pried loose and lifted in order to gain access to, and remove, the nails of the cap slates. Otherwise, they must be pulled out with a ripper.

Nails

It's often a good idea to paint ridge iron before it's installed, although new galvanized metal will not take paint unless it's first primed with a *galvanized metal primer*. When painting new ridge iron, put a coat of galvanized metal primer on first (this is not just any metal primer), then a coat of tinner's red. Neither copper nor aluminum need to be painted.

Old timers will tell you that you can prepare new galvanized metal for paint by etching it with acid first, instead of painting it with a galvanized metal primer. Some use diluted muriatic acid, while others just use a strong vinegar, and they paint it on the new metal once or twice, rinse it, dry it and then paint the metal with tinner's red. This is how I was taught to do it, but it doesn't really work very well. Better to just get some galvanized metal primer.

Always remove the old ridge iron before installing new ridge iron. It's a bad practice to install new ridge over old, as the old ridge prevents the new ridge from laying well on the roof. A common flat pry bar ("wonderbar") and 16 ounce claw hammer or CK slate hammer work well to remove old ridge iron. Straddle the ridge and pry the nails out one at a time by prying against the metal ridge, *not* against the slate.

Often, when nailing new ridge iron on old buildings, it may seem like the nails aren't hitting anything, and they may not be. When this happens, simply nail closer to the top of the ridge, instead of near the bottom edge. It's important that the nails all hit something solid, as ridge iron, being metal, is subject to expansion and contraction and the nails will work loose if not lodged into something solid.

It's also very common to find old nail holes along ridges, and it is critical when replacing ridge iron to examine the roof closely for any exposed nail holes (or heads) and to caulk them when caulking the nail heads on the new ridge. One small nail hole near a ridge will cause water

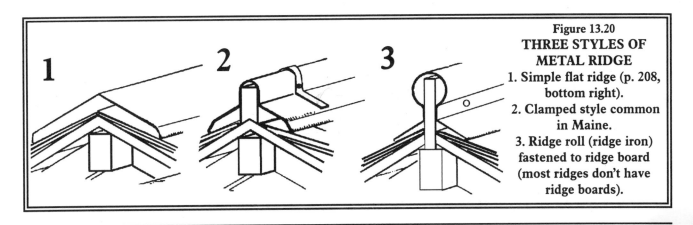

Figure 13.20
THREE STYLES OF METAL RIDGE
1. Simple flat ridge (p. 208, bottom right).
2. Clamped style common in Maine.
3. Ridge roll (ridge iron) fastened to ridge board (most ridges don't have ridge boards).

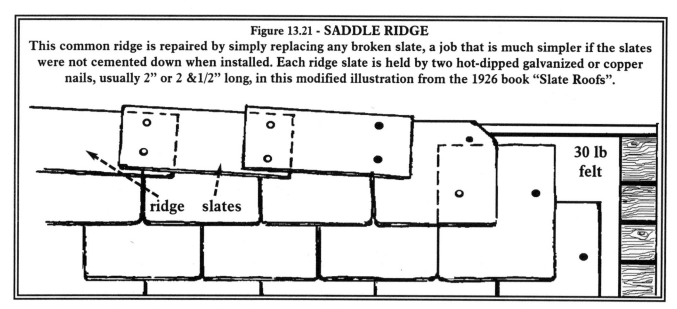

Figure 13.21 - SADDLE RIDGE
This common ridge is repaired by simply replacing any broken slate, a job that is much simpler if the slates were not cemented down when installed. Each ridge slate is held by two hot-dipped galvanized or copper nails, usually 2" or 2 &1/2" long, in this modified illustration from the 1926 book "Slate Roofs".

ridge   slates

30 lb felt

to drip through a bedroom ceiling. One dab of silicon will put a stop to it.

It's a good practice to crimp any *exposed* ends of the ridge iron (at each gable end of a gable roof, and at the bottom of each hip on a hip roof) before installing it. The end is crimped to keep out bats and bees, and to improve the appearance of the job (Figure 13.18).

Aluminum ridge metal is not recommended for use on slate roofs. Aluminum is a bit too flimsy and tends to come loose (usually because it's nailed with iron nails which react galvanically and rust). True, it doesn't need painted, and it can be nailed with aluminum nails thereby avoiding galvanic action, but aluminum ridge just doesn't hold up like a heavy gauge galvanized ridge. The ridge is one of the main "walkways" on a slate roof. Roofers will walk along the ridge to get from one place to another, and light gauge metals like aluminum will buckle under the weight of a person. This is why ventilated ridges aren't recommended on slate roofs either. The ventilation should be out the gable ends of the roof, or out roof vents, but not out the ridges. It has become a construction fad to ventilate out the ridges of roofs, but when a person works on slate roofs day in and day out, he'll cringe when he sees a slate roof with aluminum ventilated ridge, because he knows how much of a pain it will be to get around on the roof without damaging the ridge if the roof needs worked on. Sure, ventilated ridge will perform well on a slate roof, but it will interfere with the routine upkeep and maintenance of the roof by obstructing hook ladders as well as roofers. A well-done slate roof will mimic traditional styles and forego such things as ventilated or aluminum ridges, as slate roofs are made to last for centuries, and by design they should allow for routine maintenance and upkeep. People who put unusual ridges on slate roofs usually aren't thinking about having to work on the roof later.

Leaks at the ridge are often caused by exposed nail heads, both on the ridge iron and on the roof itself adjacent to the ridge. A little roof cement or silicon caulk will quickly put an end to these leaks. Additionally, the slates directly under the ridge iron sometime need replaced, and in order to do so the ridge must be lifted by prying the nails out. With the ridge lifted up a little, the nails can be removed from the cap slates using a wonderbar, claw hammer, pointed end of a slate hammer, or ripper, then the cap slates or the slates underneath them can be replaced (Figure 13.19).

Ridges that are made of slate are simply repaired by removing and replacing the slate (Figure 13.21). Some older roofing publications recommend cementing ridge slates into place (as well as nailing them). Don't use cement. In fact, it's almost never a good idea to routinely cement or caulk slates into place when installing them. Why? Because the guy who has to repair the roof years later will have a big mess on his hands when he discovers that the slates have been glued together. I know. Use roof cement or caulk only when absolutely necessary - never do it routinely. Nothing holds a slate in place better than a couple of good nails, and nails are relatively easy to remove or punch through a piece of slate when the slate is to be removed.

Old mitered hips made of slate that spread open over the years and leak can be repaired with ridge iron, as shown in Figure 12.19, page 206.

TOOLS AND MATERIALS REQUIRED FOR MOST REPAIRS AND MAINTENANCE*

TOOLS

Slate ripper; slate cutter; slate hammer (optional); 16 ounce claw hammer if no slate hammer is available; aviation snips (straight cutters will do); "wonderbar" (flat wrecking bar); small pointed trowel; 3" wide putty knife (for roof cement); paint brush, paint bucket and rag (for tinner's red); caulking gun; razor blade utility knife; nail set (1/2" bolt); ladder hooks and ladders; possibly roof jacks and planks; maybe ladder jacks. [See end of Chapter 11 for sources of tools.]

MATERIALS

Replacement slates that match in size, shape and color; pre-cut bib flashings of 16 ounce copper or painted aluminum (about 4" wide and 7" long, maybe longer or shorter); black plastic trowel grade roof cement; black fiberglass roofing membrane (a 6" roll is handy); 50 year clear silicon caulk; tinner's red oxide paint; 1 1/2" hot-dipped galvanized or copper roofing nails - (copper or brass nails needed for copper flashing); 8 penny common or galvanized nails (for galvanized ridge iron); hornet spray; 5/8" by at least 1" wooden shim strips for drip edge.

COPPER (also see Chapter 15) - Available from retailers and wholesalers including local professional building supply centers:
McClure-Johnson, 201 Corey Ave., Braddock, PA 15104; Ph: 800-232-0018 or 412-351-4300
Cassady-Pierce Co., 2295 Preble Ave., Pittsburgh, PA 15233; Ph: 412-321-8987, 800-227-7239
Copper Sales (Una-Clad), 1001 Lund Boulevard, Anoka, MN 55303; Ph: 612-576-9595
Revere Copper, 800-448-1776, Fax: 315-338-2224
May be available in smaller quantities from individual roofing companies.

OTHER METALS (Stainless, Aluminum, etc.):
Amari Metals, 2600 Freedland Road, Hermitage, PA 16148; Ph: 412-983-1900 or 800-223-9677; Fax: 412-983-1203

GALVANIZED AND COPPER RIDGE METAL - Available wholesale from:
Benjamin-Obdyke Inc., 65 Steamboat Drive, John Fritsch Industrial Park, Warminster, PA 18974-4889; Ph: 215-672-7200; Fax: 215-672-3731
Available from local professional building supply outlets and roofing companies. Insist upon 26 gauge galvanized ridge roll, or 16 ounce copper ridge roll.

SLATE HOOKS - Available from the slate supply companies listed at the ends of Chapters 11 and 12.

*Check index for sources of finials, weather vanes, lightning rods, snow guards and other accessories. Check page 185 for sources of tools and equipment, and page 218 for sources of new and used slates.

▲ Old pegged slate roof on Pembrokeshire, West Wales
▼ Old garages in North Wales. (Photos by author)

Chapter Fourteen
Flashings

Flashings are pieces of sheet metal that join the slate roof to objects that protrude through the roof, such as dormers, chimneys, pipes, or skylights, in such a manner as to prevent any water from leaking through the joint. Flashings also act as a joint between different roof planes - valleys join two planes that slope into each other, while top flashings on shed-roof dormers join two parallel planes of different slopes (Figure 14.1). Some flashings act as a joint between the roof and a wall abutting either the side, top, or bottom of the roof.

TERNE

The most commonly used flashing on old slate roofs is terne metal flashing, referred to as "tin" by some. Terne is a copper bearing steel coated on both sides with a lead-tin alloy. The typical coating weights of 20 lbs. and 40 lbs. refer to the total weight of lead-tin alloy on both sides of 112 sheets that are 20x28 inches. In fact, 20 lb. terne has .047 lbs. per square foot of coating, while 40 lb. terne has .092 lbs. per square foot. Terne metal makes a good flashing that will last for many years - 90 year old terne metal valleys on slate roofs are not uncommon, but terne metal must be kept painted or its longevity is severely reduced. In fact, unpainted terne will begin to show rust in a year or so. "Tinner's Red Oxide" is the traditional paint for painting terne on roofs, and is available at better hardware and paint stores.

Now for some good news and some bad news. The bad news is that most older slate roofs have been neglected and the flashings have deteriorated to the point that they need replacing.

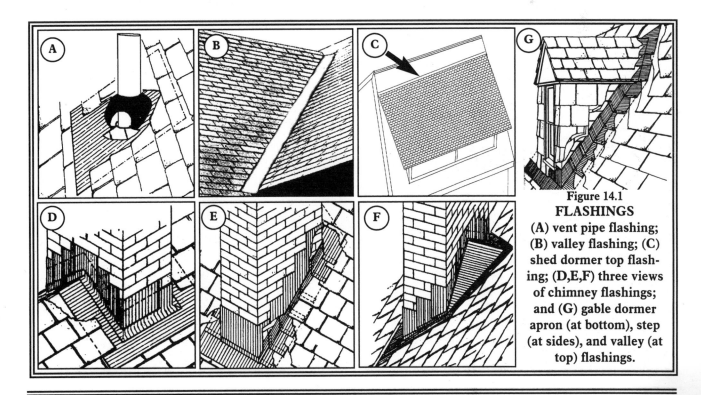

Figure 14.1
FLASHINGS
(A) vent pipe flashing;
(B) valley flashing; (C)
shed dormer top flashing; (D,E,F) three views
of chimney flashings;
and (G) gable dormer
apron (at bottom), step
(at sides), and valley (at
top) flashings.

The good news is that almost any hard slate roof can be reflashed, and in most situations no special tools or equipment are required (other than what's listed in the chapter on Tools and Equipment), except for an additional hand tool or two that is easily obtainable.

When removing terne metal flashing from slate roofs it's advisable to replace it with a metal that doesn't need painted, such as copper, stainless steel, lead coated copper, terne coated stainless, or, under some circumstances, aluminum. Terne can be used to replace terne, of course, but its use may best be limited to situations where maintenance painting will be done regularly. Experience has shown that many more roofs go neglected than are well maintained, which is the primary incentive for the use of non-corrodible metal. Yet terne metal is typically less expensive than the others mentioned above, and expense often dictates one's course of action.

STAINLESS STEEL

Stainless steel makes for an extremely durable roof flashing and can hardly be excelled for longevity. It remains shiny indefinitely, which may make it aesthetically unsuitable for some applications. Stainless falls into several categories: the 300 series typical of flashings are alloys of steel incorporating chromium, nickel, and manganese; type 302 consists of 18% chromium and 8% nickel and is most often specified for flashing, and is used interchangeably with type 304, a lower carbon variation of 302; type 316 also contains molybdenum; type 318 contains 2-3%

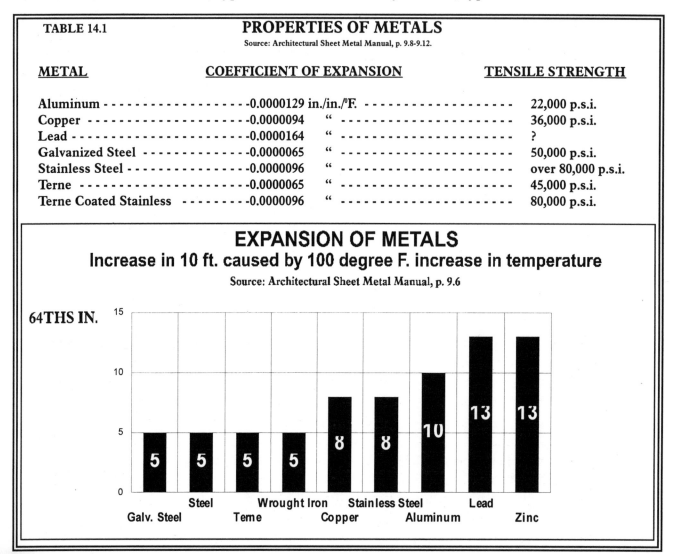

TABLE 14.1	PROPERTIES OF METALS	
	Source: Architectural Sheet Metal Manual, p. 9.8-9.12.	
METAL	**COEFFICIENT OF EXPANSION**	**TENSILE STRENGTH**
Aluminum	-0.0000129 in./in./°F.	22,000 p.s.i.
Copper	-0.0000094 "	36,000 p.s.i.
Lead	-0.0000164 "	?
Galvanized Steel	-0.0000065 "	50,000 p.s.i.
Stainless Steel	-0.0000096 "	over 80,000 p.s.i.
Terne	-0.0000065 "	45,000 p.s.i.
Terne Coated Stainless	-0.0000096 "	80,000 p.s.i.

EXPANSION OF METALS
Increase in 10 ft. caused by 100 degree F. increase in temperature
Source: Architectural Sheet Metal Manual, p. 9.6

64THS IN.

Galv. Steel 5, Steel 5, Terne 5, Wrought Iron 5, Copper 8, Stainless Steel 8, Aluminum 10, Lead 13, Zinc 13

molybdenum and is more corrosion resistant than the other types of stainless steel; the 400 series contains no nickel and is therefore less resistant to corrosion and is used in interior applications.

Stainless steel used for detailed flashings is best fully annealed and dead soft, while harder stainless steel makes excellent valley flashing on slate roofs. Stainless also comes in a variety of finishes ranging from non-reflective to a mirror finish.

Terne-coated stainless is type 304 stainless coated with terne, and it's another popular flashing material. The terne coating eliminates the shiny surface gloss of pure stainless, and it makes the normally difficult stainless much easier to solder. Prolonged handling of this material with bare hands may pose a health risk.

LEAD

Sheet lead is a very popular flashing material in Europe, and is used there for valleys, ridges, and just about everything else on slate and tile roofs. Again, the prolonged handling of lead can pose a health risk to the people using it. Lead poisoning, caused by the absorption of lead into the body, can result in anemia and paralysis, so gloves should be worn when handling lead or lead-bearing metals.

Sheet lead, like copper, is specified according to the weight per square foot, and is available in grades ranging from one to eight pounds per square foot. Four pounds per square foot of sheet lead yields a sheet that is 1/16 inch thick.

One positive characteristic of lead flashing is its workability - it bends easily and can be made to conform to irregular surfaces. It's also very durable and highly resistant to atmospheric corrosion, and will not stain a roof like rusting terne or galvanized will. A drawback to lead is its high expansion rate, and allowances must be made for expansion and contraction of lead when used in larger sheets.

GALVANIZED

Galvanized metal flashing has been commonly used in roofing applications. It is composed of steel or iron coated with zinc on both sides, either hot-dipped or electro-plated. It's fairly inexpensive and long-lasting if kept painted, although it will rust quickly if neglected, and is therefore not recommended for slate roofs. Galvanized metal reacts galvanically with copper and the two metals should not be allowed to contact each other in the presence of moisture.

ALUMINUM

Aluminum is perhaps the most popular flashing material in America today for use on asphalt roofs. It comes in its natural silver-gray color (mill finish), or coated with paint (painted aluminum). The material is readily available at most building supply centers because it's used for soffit, fascia, and siding, as well as flashing, and can be bought in 50 foot rolls ("coil-stock"). It is light in weight, the thinner gauges are easily bendable, and it is highly resistant to corrosion due to the protective oxide that forms on the surface of the metal. Much of the mill finish aluminum available in rolls is too thin (.015") for any use on a slate roof, and even the painted aluminum (.019"/.020") should be avoided for any long term flashing jobs when copper is available. Painted aluminum does, however, make excellent bib flashings for covering the nails of replacement slates (see Basic Repairs chapter). Heavier gauge aluminum (.032" - .040") will make a suitable valley on a slate roof, although care must be taken to use aluminum or stainless steel nails whenever nailing aluminum valleys into place.

COPPER

Copper is a non-magnetic, malleable, corrosion resistant metal which shines like brass when new, but soon turns reddish brown upon exposure to the elements, and finally green. The most common flashing material recommended today for use on slate roofs is 16 ounce copper, which gets its name from the fact that it weighs 16 ounces per square foot. An even more durable copper flashing is 20 ounce, which, of course, weighs 20 ounces per square foot, although the 16 ounce is more workable with hand tools when detailed work is required.

Copper has long been the roof flashing of choice in America because it's durable without needing painted; it develops a green patina with age that many people find aesthetically pleasing, and which is said to form a protection against corrosion; and most importantly perhaps, it's easy to work with. Sixteen ounce copper is easy to bend, and can be bent by hand in most cases, or with a pair of sheet metal tongs. Additionally, pure copper doesn't present the toxic hazards that lead-bearing metals such as sheet lead, lead-coated copper, terne, and terne-coated stainless do, exposure to which can have an adverse effect upon the health of the people who handle it for prolonged periods. The mining of copper certainly does its share of environmental degradation as well, yet roof flashing must rank as one of the more honorable uses of the metal, and surely its costs are justified.

Copper, like all metals, does benefit from paint, which will prolong its life indefinitely. It must also be nailed with copper or brass nails, as galvanized nails will corrode in contact with copper flashing. Copper reacts galvanically with aluminum, steel, zinc, and galvanized steel, and should not be placed in contact with these metals in the presence of moisture.

Copper flashing may come as either "sheet" or "strip" - sheet copper according to the Copper Development Association in England (1951) being of exact length, over .0006 inches thick, but not over 3/8" thick, and over 18" wide; while strip copper falls into the same thickness range but may be of any width and usually comes in coils.

Copper also comes "hard," "half-hard," or "soft." Hardened copper is used, for example, in copper nails. Half-hard copper is used in situations where the copper requires a degree of rigidity, such as in copper gutters, and half-hard copper works very well for valley flashing. Soft copper is the choice for most other slate roof flashings (such as chimney flashings). Copper can be softened by annealing, or heating to a dull red with a propane torch and either quenching in water or allowing to cool naturally in the atmosphere.

Lead-coated copper is copper coated on both sides with lead, and is popular among architects because it remains dull gray in color, is durable, and is easy to solder. Prolonged use of this material may pose a health risk to the person handling it.

GALVANIC CORROSION

"Galvanic action," simply stated, is the corrosive reaction between incompatible metals. The corrosion is expedited by an "electrolyte," which is a non-metal substance which will conduct an electric current - water, for example, makes a good electrolyte, and salt water is even better. Metals that are more electro-positive (anodic) corrode more easily, and metals that are more electro-negative (cathodic) are more corrosion resistant. Common metals range from electro-positive to electro-negative according to the following scale:

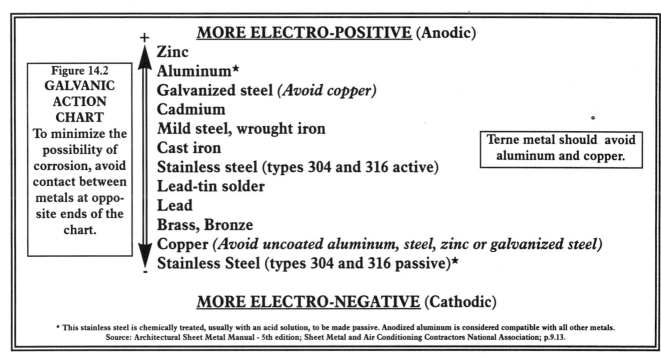

Figure 14.2
GALVANIC ACTION CHART
To minimize the possibility of corrosion, avoid contact between metals at opposite ends of the chart.

MORE ELECTRO-POSITIVE (Anodic)

\+

Zinc
Aluminum★
Galvanized steel (*Avoid copper*)
Cadmium
Mild steel, wrought iron
Cast iron
Stainless steel (types 304 and 316 active)
Lead-tin solder
Lead
Brass, Bronze
Copper (*Avoid uncoated aluminum, steel, zinc or galvanized steel*)
Stainless Steel (types 304 and 316 passive)★

MORE ELECTRO-NEGATIVE (Cathodic)

Terne metal should avoid aluminum and copper.

★ This stainless steel is chemically treated, usually with an acid solution, to be made passive. Anodized aluminum is considered compatible with all other metals.
Source: Architectural Sheet Metal Manual - 5th edition; Sheet Metal and Air Conditioning Contractors National Association; p.9.13.

The farther apart metals are on the above scale, the more they will react galvanically. For example, zinc and aluminum will react with copper much more than they will with each other. Therefore, it's important to avoid using incompatible metals in contact with each other when installing flashing. If such contact seems unavoidable, place a layer of roofing paper, roof cement, paint, or other non-absorbent, non-conductive material between the metals.

USING COPPER

According to the Copper Development Association (CDA) in England, *"In dealing with copper sheet or strip it must be realized that while the technique to be employed calls for no special skill or training, it is important that a little extra care and forethought should be given to setting out the work; this will amply repay the craftsman, and a knowledge of the physical properties of the metal will enable him to produce work equal to that of the old craftsmen whose efforts were renowned. The aim of the sheet copper worker must be to achieve the necessary shape or form with as little actual working of the metal as possible. Thus it will be seen that the working techniques of other metals is not always applicable."*

Copper expands and contracts under varying temperature conditions, but the amount of movement that takes place for a given temperature variation is about 40% less than with either lead or zinc. Nevertheless, the expansion and contraction of copper creates limitations in its use, and the Copper Development Association warns, *"It cannot be too strongly emphasized that the superficial area of each individual piece of copper sheet or strip must not be greater than 14 square feet for thicknesses up to and including .022 inches . . ."* When large sheets are used in flashing situations, the copper sheets must be fastened to the roof with copper cleats to allow for some movement of the metal.

Furthermore, *"the use of solder in connection with sheet copper work is to be deprecated and should be avoided,"* according to the CDA. *"It will only give a false sense of security, and if the use of copper*

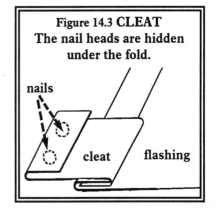

Figure 14.3 **CLEAT**
The nail heads are hidden under the fold.

nails

cleat flashing

for flashings is doubtful about any welt [folded joint] being watertight it would be better to remake it rather than try patching it with solder. The reason for this is that solder and copper have different ratios of expansion and contraction, and sooner or later this will cause the solder to fatigue and break down, thereby allowing moisture to enter. These remarks apply only to the use of ordinary fine and plumbing solders, and not to brazed joints, which in some cases may be used instead of welding [soldering]." The Architectural Sheet Metal Manual (1993) however, recommends the use of 50/50 solder (50% lead and 50% tin) when soldering copper, and recommends that the flux be neutralized (flushed with water) after soldering. Fortunately, almost no flashing jobs on typical slate roofs will require soldering of any kind, nevertheless a section on soldering is included later in this chapter.

BASIC FLASHING PRINCIPLES

Older slate roofs develop two primary flashing problems: deteriorated valley flashing and deteriorated chimney flashing. A third, less serious, but still common flashing problem involves deteriorated sewer vent pipe flashings. Other flashings that go bad on slate roofs include dormer step flashings (sides of dormers), dormer top flashings on shed roof dormers (Figure 14.1), skylight flashing, and flashings that connect the slate roof to other roofs, or to the outside walls of the building.

Flashing a Chimney

Imagine, if you will, a flat plane such as a floor, and through the plane protrudes a rectangular column such as a chimney. The four sides of the column can be easily flashed by simply bending a piece of metal to a ninety degree angle and fitting it against the column and against the flat plane (Figure 14.4). If the column were an actual chimney, the top edges of the flashing would be bent and fit into a mortar joint to prevent water getting behind the flashing.

What's wrong with this picture? Two things: a) on a roof, the plane is not flat, it's sloped, and b) the four corners of the column will leak because they're not flashed, only the sides are flashed. Corner leakage is the single most important problem people encounter when trying to flash objects protruding through a roof. The solution is simple, however, and involves no soldering or special sheet metal work. But first, let's look at the same scenario with a sloped plane (Figure 14.5). Now we have an advantage, because the sloping of the plane forces water to flow in one direction, and because we know the direction of flow, we can control the water, and therefore eliminate any leaking.

The sloping plane also adds a new factor into the situation: the top of the flashing is no longer parallel to the mortar joints and cannot be folded into a joint as is. Instead, the top of the flashing must be cut to follow the stepped joints, and folded in accordingly (Figure 14.6).

Next, we must take into consideration that several rows of slate are going to butt against the side of the chimney. Since the slates overlap each other, the side flashing on the chimney must be cut into sections that overlap each other as well, because one long piece of side flashing may allow water to run

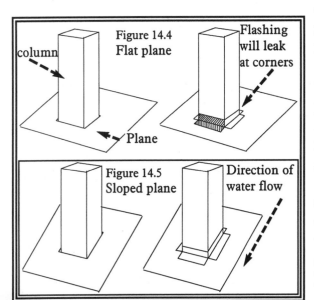

Figure 14.4
Flat plane

column

Flashing will leak at corners

Plane

Figure 14.5
Sloped plane

Direction of water flow

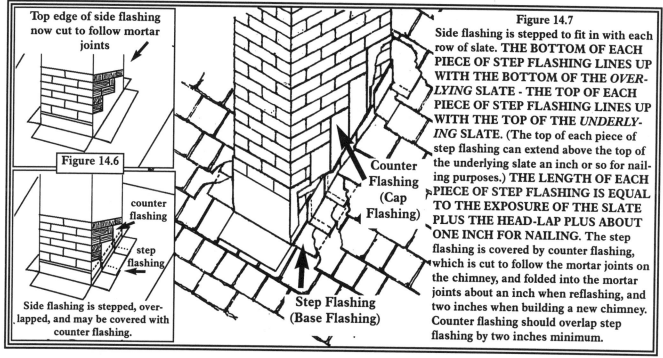

Top edge of side flashing now cut to follow mortar joints

Figure 14.6

Side flashing is stepped, overlapped, and may be covered with counter flashing.

counter flashing

step flashing

Figure 14.7
Side flashing is stepped to fit in with each row of slate. THE BOTTOM OF EACH PIECE OF STEP FLASHING LINES UP WITH THE BOTTOM OF THE *OVERLYING* SLATE - THE TOP OF EACH PIECE OF STEP FLASHING LINES UP WITH THE TOP OF THE *UNDERLYING* SLATE. (The top of each piece of step flashing can extend above the top of the underlying slate an inch or so for nailing purposes.) THE LENGTH OF EACH PIECE OF STEP FLASHING IS EQUAL TO THE EXPOSURE OF THE SLATE PLUS THE HEAD-LAP PLUS ABOUT ONE INCH FOR NAILING. The step flashing is covered by counter flashing, which is cut to follow the mortar joints on the chimney, and folded into the mortar joints about an inch when reflashing, and two inches when building a new chimney. Counter flashing should overlap step flashing by two inches minimum.

Counter Flashing (Cap Flashing)

Step Flashing (Base Flashing)

under the edge of the slates. The flashing is then installed in overlapping sections and "stepped" up the side of the chimney, each step lining up with a row of slates. This is known as "step flashing." To make step flashing easier to install, it is often not folded into the mortar joints at all, but instead it's just run up the side of the chimney two or three inches, then covered with "counter flashing," which is also installed in overlapping sections (it overlaps itself as well as the step flashing). Step flashing is also known as "base flashing," and counter flashing is also known as "cap flashing" (Figure 14.7).

The side flashing pieces need not be installed as separate step and counter flashings, but can be installed in separate overlapping pieces that are both stepped over each other *and* folded into the mortar joints (Figure 14.8). This single side-piece system works well on smaller chimneys, while very large brick chimneys and stone chimneys may be easier to flash using separate step and counter flashings.

Figure 14.8
Single piece side flashing - step and counter flashing combined in one piece.

We have now demonstrated how to flash the *sides* of a column protruding through a sloped plane, but what about the corners? What about the top side and bottom side of the column? Well, all four corners are rendered leakproof by building little "roofs" over them with the flashing. The bottom corners are covered with the bottom piece of side flashing and the top corners are covered with the uppermost section of chimney flashing (Figure 14.9). In both cases this is simply achieved by extending the flashing beyond the corner about four inches, then folding the flashing around the corner and tucking it into the mortar joint. Extending the flashing beyond the corner forces the water to flow away from the corner, while folding the flashing around the corner and into a mortar joint prevents water running down the outside of the chimney from finding its way into the

TOP FLASHING

The top flashing must extend beyond the chimney corner about 4", and should be made from 24" wide 16 ounce copper stock. This flashing must extend as far up the roof under the slate as the width of the flashing stock will allow, and must be folded against the base of the chimney tightly.

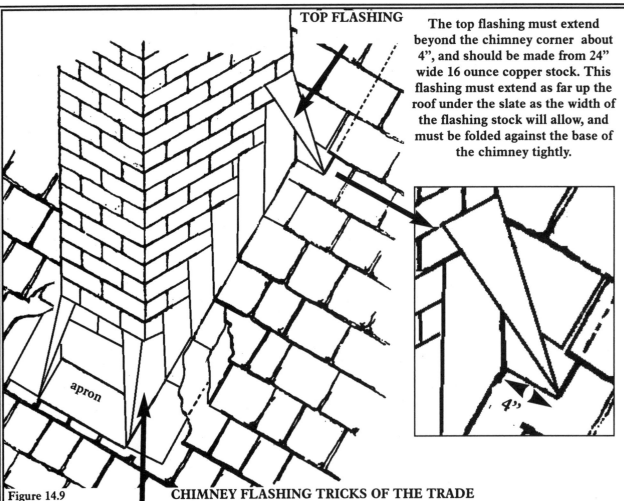

apron

Figure 14.9

CHIMNEY FLASHING TRICKS OF THE TRADE

The lowest piece of side flashing is extended below the bottom corner of the chimney about four inches, then folded around the corner and tucked into the mortar joint. A couple of inches of excess flashing must be allowed for at the top edge of the side flashing in order to have enough material to fold around the corner and still reach the mortar joint. The corners of the chimney can be caulked to keep out insects, but no water will get in due to the flashing "roof" that protects it. The top chimney corners are protected in the same manner, by projecting the top flashing beyond the corner about four inches, then folding the flashing around the corner and tucking it into a mortar joint. Again, excess metal must be allowed for in order to have enough flashing to reach the mortar joint. The top piece of flashing on a chimney is very important as snow and water dam up against the back of the chimney. When fitting this flashing into place, make sure it fits tightly down against the base of the chimney where the chimney meets the roof (make sure your flashing fold at that point is sharp). When building a new chimney allow two inches of the flashing to fold into the mortar joints. When reflashing an old chimney, you'll be lucky to get an inch into the mortar joints, and you should be content with 3/4". Fasten the flashing in the mortar joint of an old chimney by wedging a 1 1/2" x 2" folded piece of flashing metal into the joint to hold the flashing (see below), then caulk all the flashed joints, when done, with 50-year silicon. Figure 14.14 (p. 249) illustrates a reflashing sequence on an actual chimney.

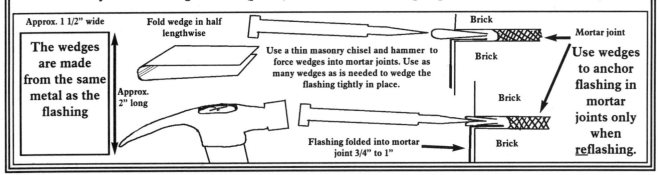

Approx. 1 1/2" wide

The wedges are made from the same metal as the flashing

Approx. 2" long

Fold wedge in half lengthwise

Use a thin masonry chisel and hammer to force wedges into mortar joints. Use as many wedges as is needed to wedge the flashing tightly in place.

Flashing folded into mortar joint 3/4" to 1"

Brick

Mortar joint

Brick

Brick

Brick

Use wedges to anchor flashing in mortar joints only when **reflashing.**

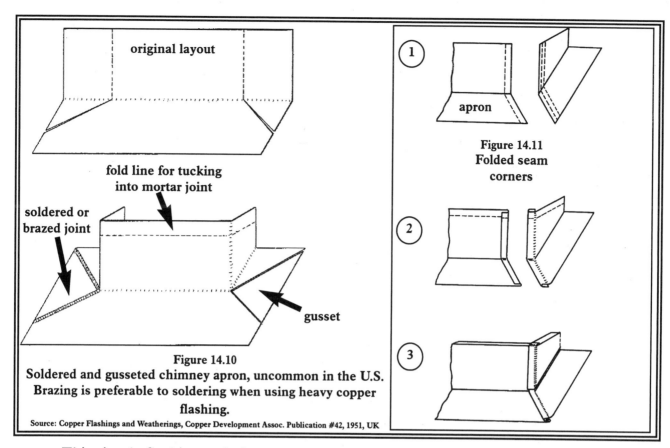

original layout

fold line for tucking
into mortar joint

soldered or
brazed joint

gusset

Figure 14.10
Soldered and gusseted chimney apron, uncommon in the U.S.
Brazing is preferable to soldering when using heavy copper
flashing.

Source: Copper Flashings and Weatherings, Copper Development Assoc. Publication #42, 1951, UK

1

apron

Figure 14.11
Folded seam
corners

2

3

corner. This simple flashing technique can be used for any rectangle protruding through the roof, such as a skylight, and it requires no soldering, adhesives, roof cement, caulk, or fancy sheet metal work. Most importantly, if done correctly, it's foolproof. The size of the "roofs" built over each corner is variable and may be made larger or smaller according to the situation, or according to personal preference.

An alternative to the folded-corner method of flashing involves fitting the apron flashing (the lowest piece on the chimney) with soldered gusset pieces, thereby eliminating any holes in the lower corners (Figure 14.10). This style of flashing was once popular among English roofers who were involved in a lot of sheet metal work, and to whom the soldering or brazing of flashing was part of the normal routine. Another technique that had some degree of popularity was the use of folded seams at the corners of the chimneys to provide a waterproof flashing job (Figure 14.11). This technique also was limited to the domain of sheet metal specialist.

The apron flashing on the chimney is the piece that fits at the lowermost front of the chimney. Often this is seen installed as two pieces, a base piece and a cap piece; however, one piece is sufficient and easier to install. The apron should extend down the roof far enough to cover any exposed nail heads, as well as to provide a headlap on the second course of slate below it. Four to six inches of flashing extending down the roof is common. The apron need not extend beyond the sides of the chimney, although it can be allowed to do so and then can be folded against the roof alongside the chimney to allow a place to nail it (Figure 14.12).

After the apron is installed, the side pieces are installed. They should lay on the roof about four inches and run up the side of the chimney 2-4" for step and counter flashings, and until they reach and tuck into a mortar joint for single piece flashings (see Figure 14.8 and 14.9 for single piece side flashings, and 14.7 for step and counter flashing). Single piece side flashings should not be cut on the ground based on drawings and measurements, but instead should be

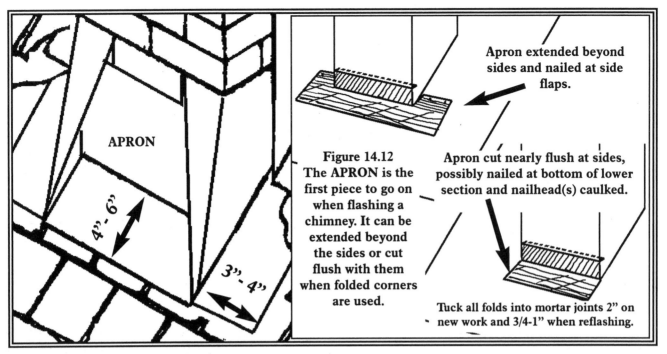

APRON

4" - 6"

3" - 4"

Apron extended beyond sides and nailed at side flaps.

Figure 14.12 The APRON is the first piece to go on when flashing a chimney. It can be extended beyond the sides or cut flush with them when folded corners are used.

Apron cut nearly flush at sides, possibly nailed at bottom of lower section and nailhead(s) caulked.

Tuck all folds into mortar joints 2" on new work and 3/4-1" when reflashing.

laid against the chimney with the 4" fold against the roof and marked in place using a level and a pencil, then cut and installed (Figure 14.15). This ensures a good, tight fit, both against the chimney and against the roof. If step flashings and counter flashings are used, the *bottom* edges of the counter flashings should be folded under about half an inch for stiffness.

The uppermost piece of flashing on a chimney is very important as it catches the brunt of the rain and snow. A wide stock of copper should be used; 24" is usually sufficient, and need not be run up the back of the chimney more than eight inches (three courses of bricks). That leaves about 16" or more to run up the roof, and care should be taken to not nail through the roof section of flashing when laying the slate over it, except within one inch of the top edge. The slate should not be laid tightly against the back of the chimney, but the flashing on the roof at the base of the chimney should be left bare for at least a few inches to allow the water to quickly run off. It's very important that the flashing fit tightly against the base of the chimney, otherwise capillary action may allow water to creep under the side wings and make its way into the chimney cor-

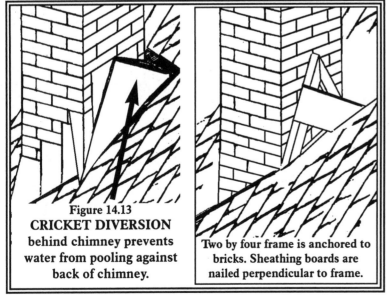

Figure 14.13 CRICKET DIVERSION behind chimney prevents water from pooling against back of chimney.

Two by four frame is anchored to bricks. Sheathing boards are nailed perpendicular to frame.

ners. A tight fit can be achieved by over-bending the flashing to make a tight crease, then opening it back up to its proper position before laying it in place.

The top flashing should extend beyond the sides of the chimney about 4" and be folded around the chimney corner and tucked into a mortar joint as shown in Figure 14.9. In order to do this, one must allow some extra flashing length where the metal will tuck into the mortar joint on the back of the chimney, so if you're going to tuck 3/4" into the mortar joint, leave 2" of metal instead. The extra length is trimmed off with a pair of tin snips *only* where the flashing tucks into the back side of the chimney, but <u>not</u> where it extends beyond the sides. This should leave enough flashing material to reach around the corner and tuck into the side mortar joint. This principle is illustrated in Figure 14.15, steps 7, 8, and 9.

There are two places where hidden leaks may occur after installing the piece of top flashing on a chimney. The first can occur at the top edge of the flashing, which is under the slate, but may be exposed in the slots. If so, slide extra pieces of flashing under the slots but over the metal to cover the exposed edges. Secondly, the side edges of the flashing (where the water runs off) may fall on or near a slot between the slates. If so, reinforce the slot with flashing slid underneath the slates.

Finally, when a chimney is very wide and creates a large dam on a roof and therefore a potentially chronic problem, a "cricket" should be installed between the top side of the chimney and the roof (Figure 14.13).

Practice makes perfect. Flashing work is a skill that benefits from practice, and the more one does of it, the easier it gets. Nevertheless, a person with no prior experience should be able to flash or reflash a chimney, although they may have to take their time and pay close attention

Figure 14.14
COMMON CHIMNEY REFLASHING JOB - BEFORE AND AFTER
Single flue chimneys exiting through the ridge of a roof are common, and unfortunately so are the ugly messes that are evident in the photo at the left. In about five hours a single worker cleaned up the roof and reflashed the chimney, as shown at right. The step by step procedure is illustrated in Figure 14.15.

Figure 14.15: ▲ Step 1 - Remove all old flashing and bad slates.

Step 2 - ▲ Replace bad slates, but leave caps off for now.

Step 3 - ▲ Grind out old mortar joints with grinder in preparation for new flashing.

Step 4 - ▲ Install apron, with wedges. Step 7 - ▼ Cut flashing for side joint, leave excess to cut later.

Step 5 - ▲ Lay side flashing against chimney, 4" bend on roof. Step 8 - ▼ Fold side into mortar joint.

Step 6- ▲ Mark cut lines. Step 9 - ▼ Fold around corner, trim and tuck into joint. Replace remaining slate. Add ridge.

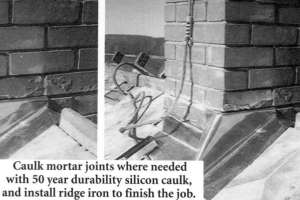

Caulk mortar joints where needed with 50 year durability silicon caulk, and install ridge iron to finish the job.

to detail to get it just right. Tools required for most flashing work include a pair of tin-snips (straight cutting aviation snips), sheet metal tongs for bending the edges that fit into the mortar joints, a thin bladed masonry chisel for cleaning out mortar joints and for forcing in wedges to hold the flashing (reflashing only), and a two-foot level. An electric grinder with a masonry wheel or diamond blade is very useful for cleaning out old mortar joints in preparation for reflashing. Figure 14.15 shows a chimney reflashing job done in step-by-step sequence.

VALLEYS

Worn out valleys on old slate roofs create one of the biggest problems any roof owner could have. Valleys collect water from two or more roof planes and therefore channel a greater concentration of water than any other part of the roof. A hole the diameter of a match stick in a valley can cause buckets of water to leak into a building during heavy rains. Worse, the bottom ends of the valleys often wear out first, and they also bear the greatest amount of water, so serious structural water damage can happen to the area at the bottom of the valley if left leaking and neglected for a long time.

Fortunately, valleys aren't difficult to replace, and valley replacement is a routine job when restoring old slate roofs. When valleys leak they should be replaced in their entirety. That means the old valley flashing must be completely removed and new valley flashing installed in its place. Some roofing contractors will try to slip sections of flashing over the old metal without removing the old valley. This doesn't work. Others will tar the old valley and often the roof adjacent to the valley on both sides. This is a mistake. Valleys can be sealed as a temporary measure to prevent leaking until they're replaced, and the procedure for doing so is illustrated in Chapter 13, page 226. But if the job is to be done permanently, correctly, and only once, the old valley must be removed and a new one installed.

In order to remove an old valley one must simply remove enough slates on both side of the valley to expose the old valley metal, which is usually about 14"- 20" wide overall. The metal is then pried loose and removed, the exposed sheathing underneath swept clean with a broom, a

Figure 14.16
A standard, open, 16 ounce copper valley on new construction, with recycled slate. Lines are drawn on the 16 inch wide copper with a permanent felt tipped pen (following chalk lines) to align the slates along the 6" wide exposed valley edge.

line is chalked for the new valley, the new metal is nailed in place, and the original slates are replaced. If the slate along the valley is in good shape, a competent slater can expect to replace 34' of valley in a 10 hour day, working alone. If the slate is not good (i.e. tarred and/or broken), a competent slater and a competent helper can expect to replace a 34' valley in a long (10 or 12 hour) day. The slates along a valley are often not good because people who have attempted to repair the valley in the past have walked all over it, thereby damaging the slate, and they've spread tar all over the valley and adjacent slates, thereby making it very difficult to take the roof apart. It's important to replace a valley in a single day, however, as a roof should not be torn apart and left open and subject to a potentially disastrous cloud burst.

Taking the old Valley Out

The single most important step in removing old valleys is the roof set up. Typically, a hook ladder is positioned on either side of the valley, set back far enough to allow access to the slates that must be removed (about 3-4" back from the valley center). If the roof is steep, the hook ladders will try to slide into the valley and must be tied to a roof jack nailed to the roof for that purpose. If the valley is long, such as the 34' length referred to above, a 20' hook ladder (which is the longest practical hook ladder available) will be too short to extend the entire length of the valley.

Figure 14.17
The slates are removed from either side of this 20' valley on a house in Sandy Lake, PA, and propped on the hook ladder, which is prevented from sliding into the valley by a roof jack tied to the ladder.

Figure 14.18
After the new metal is installed (in this case 20 ounce copper), the numbered slates are nailed back in place, and any bad slates are replaced in the process.
(Both photos by the author)

Then two hook ladders must be hooked together, the lower, shorter hook is reversed to face away from the roof so it can hook on a lower rung of the upper, 20' hook ladder. A long valley on a steep roof may require more than one roof jack to anchor the two ladders in place and prevent them from sliding into the valley.

And they will slide into the valley, because the slates, when removed, are propped on the hook ladders as they come off the roof, and this weighs the ladder down. It also, however, makes it very easy to put the slates back on the roof after the new valley metal has been installed because the slates are waiting adjacent to where they were removed.

The trick-of-the-trade that makes it even easier to put the roof back together is *numbering* the slates before they're removed. Starting at the top row, a nail or other sharp instrument is used to scratch the number "1" on the first row, the number being scratched on the first 4 or 5 slates out from the valley, on both sides of the valley. Then the second row is scratched "2," the third row "3" and so on, until the entire length of the valley is marked, row by row, about 4 or 5 slates out from the valley center (Figure 14.19A). Mark the starter slates at the bottom with an "ST". This only takes a few minutes, but saves an incredible amount of time when the valley must be reslated. Measure the exposed width of the valley and make a note of it before beginning to remove the slates.

Pry the ridge iron, if any, loose enough to remove the nails from the cap slates underneath. The numbered slates are then removed, starting at the top, using a slate ripper, flat pry bar (mostly), and hammer, and the slates are propped on the hook ladders as they're removed. *All the slates from the right side of the valley must be propped on the right hook ladder, and all the slates on the left side of the valley must be propped on the left hook ladder.* Do not overlook this important detail if you want to put the valley back together quickly and efficiently. Also, remove the slates from one side of the valley at a time for maximum speed and efficiency (expose one entire side of the valley before exposing the other).

If the slates have been tarred into place, try to carefully pry them loose without breaking them. In some cases you may have to cut along the tarred valley edge with a razor utility knife to get the slates to come loose in one piece. Heavily tarred slates that can be pried loose can usually be cleaned of most of the tar and re-used. The tar is knocked off by tapping directly against the edge of the slate with a hammer, or by using the tip of a slate ripper or flat pry bar to chip it off.

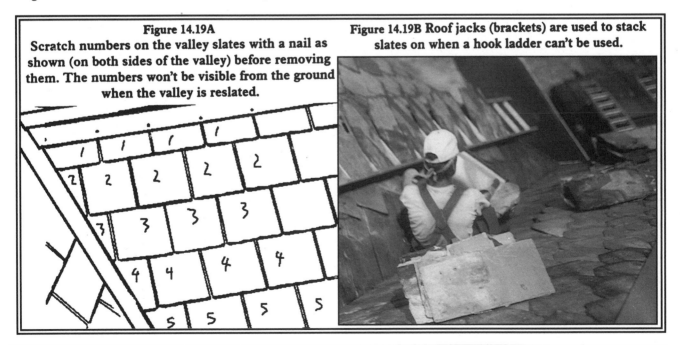

Figure 14.19A
Scratch numbers on the valley slates with a nail as shown (on both sides of the valley) before removing them. The numbers won't be visible from the ground when the valley is reslated.

Figure 14.19B Roof jacks (brackets) are used to stack slates on when a hook ladder can't be used.

If you cannot place a hook ladder on either side of the valley because there isn't a ridge to hook onto, then you must nail roof jacks to the roof and prop the slates on them as in Figure 14.19B (see page 181 for instructions on nailing roofs jacks). Again, the slates from one side of the valley must be propped on that side of the roof, and vice versa.

Once all the slates are removed and the old valley flashing is completely exposed, the old metal can be pried out with that flat pry bar. Now is a good time to warn you that this is a seriously dirty job, and a dust mask is a good idea when taking out old valleys. Also, be on the lookout for wasp nests underneath the bottom of the old valley when you roll it out. Pry the old metal loose, roll it down the roof starting at the top, and drop it over the edge, if possible. Then sweep the filthy roof clean with a broom, and examine it closely for nails and any sharp objects that might rub against the new flashing and eventually penetrate it. Use a gloved hand and brush it over the roof surface. It will catch on any sharp nub. Pound the nub in or pull it out with a claw hammer. When you're sure you have the old sheathing clean, chalk a line on one side of it the length of the valley to indicate where the edge of the new flashing will be. If the new metal is 16" wide (recommended), snap your chalk line at 8" from the center of the valley.

Often there will be a small spot on the sheathing that has rotted due to a prolonged leak. If the new flashing is going to cover it, don't worry about it - put the new metal in right over top of it. It's very rarely that the sheathing in a valley will be so bad that carpentry work needs to be done. It's more important that the valley be replaced and the roof be closed up as quickly as possible, so minor imperfections in the valley sheathing can be tolerated. They'll do no harm.

Installing the New Valley Flashing

When laying the new metal in place, remember that it doesn't have to be creased down the center on a sheet metal brake, nor does it have to be "cleated" in place (fastened with cleats - refer to page 243 to refresh your memory on this issue if necessary). Instead, the metal simply should be carefully forced into the contour of the valley into a rounded valley shape. This won't work, however, if the *exposed* valley is too *narrow* (under 4") on a *steep* roof - then the valley needs creased. However, narrow valleys on steep roofs are uncommon, and ninety-nine times out of a hundred a force-fit valley works beautifully.

It's important to remember that the valley metal should not be laid in lengths longer than about ten feet due to possible expansion and contraction of the metal and consequent long-term harm. Twelve foot long valleys may be necessary in some instances but should be the limit. The sections should be overlapped six inches in the direction of water flow. One side of the valley should be nailed first along its entire length (the chalked side), then one end forced into place (usually with a knee) and nailed, then the other. The end that you want to fit tightest (usually the bottom end of an *overlapped* joint because that overlapping is visible from the ground) should be nailed last. Be very careful when forcing a copper valley into place as too much force can cause the soft metal to kink. However, get the valley metal to conform to the shape of the roof as tightly as possible without improperly bending it.

When replacing two valleys that converge at the top (a common situation), fold the tops over each other at the ridge and remember that there will probably be a piece of ridge that will cover up some of that valley top and keep the rain out. What doesn't fold over the ridge will lay flat on the roof and should also be overlapped. If the overlap has not been done adequately, an extra piece of metal can be placed over it to make it leak proof before the roof is re-slated. Practice makes perfect.

The bottom of the valley usually has to be cut to conform to the corner of the roof, leaving

▲ Hook ladders are positioned alongside this 34' long church valley. The valley was tarred, but the sea green slate was quite good.

▲ The slates are pried off, stacked on the hook ladders and the old valley metal removed. The roof is swept and chalked.

▲ A 12' section of .025" stainless steel, 16" wide, is forced into place and nailed along the edges, aligned with the chalk line.

▲ A second section of stainless over laps the first by 6".

Figure 14.20

▲ The third and final section of stainless is installed.

The slates are nailed back where they came from. The new valley should last the life of the roof and require no maintenance.

enough overhang to channel the water into the gutter. If you're not sure how much overhang to leave, better to leave too much - it can be cut off later. Make sure the valley metal is *under* the shim strip at the drip edge and not on top of it (assuming a shim strip exists). Keep the shim well back from the valley center - it need only overlap the valley metal an inch or two.

Next, chalk two lines up the valley to indicate where the edges of the slate will be. Some foresight is necessary here, and although most valleys have parallel exposed edges, not all exposed valleys are the same width. Six inches is average, but they'll vary from 4" to 8" if open, but may be closed instead. If the original valley exposure was 6" wide, measure 3" from the center on both sides and chalk two lines the length of the valley on the metal, 6" apart. Then imme-

Figure 14.21
A carefully placed knee will force the valley metal into position without buckling it (above left). After nailing the valley in place, one side is re-slated first while working off the bare roof (above, middle). Then the hook ladder (now empty) on that side can be moved in close to the valley to sit on while slating the other side, thereby avoiding damaging the newly replaced slate.

diately go over those lines with a permanent felt tip marker pushing the chalk out of the way with the felt tip as you go (the ink will eventually wear off). When re-slating, align *the inside edges of the slates with those ink lines* and don't worry if the *sides* of the slates are exactly parallel if you want the *valley* to look neat. If you're satisfied with a somewhat irregular valley more closely resembling the original, then don't bother chalking and drawing the lines at all, just line the slates up with the original weather marks on the slates themselves as you reslate the valley.

Reslating

When re-slating the roof, start at the bottom. Use the same slates that were removed from the roof as much as possible, for two reasons. First, they'll match that roof better than any other slates because they came from that roof. Secondly, they have weather marks on them showing where they were overlapped by the other slates, and can therefore be placed back in the same spot they came from by simply lining up the weather marks with the slates already on the roof. With numbered slates positioned on the proper side of the valley, and weather marks to guide you, you can zip a valley back together in less time than you may think.

When you start re-slating at the bottom of the valley, make sure you have a piece of wood shim in place under the starter slates, about 1/2"- 5/8" thick and an inch or so wide, positioned right at the bottom of the sheathing (see page 201). If the original one is still there and is usable, fine. The shim strip lays *on top* of the valley metal, but should not extend into the center of the valley as far as the slate, and usually *need not extend over the valley metal very far at all.*

When re-slating, simply nail the slates back into place, in the same spots they were removed from - *one side of the valley at a time*. After one side has been reslated, slide the hook ladder from the finished side closer to the valley and work off it while slating the other side. Be careful that the hook ladder is not digging into the valley metal (Figure 14.21).

You'll find that many of the slates you're re-nailing will be covered on one side by another slate and therefore must be nailed only on one side. Punch an extra hole on the one exposed side with a slate hammer and nail every slate with two nails. If the old nail holes are still good, use them too, but make sure your nails are 1 1/2 inches long to get that extra bite (most original nails were 1 1/4" long). Occasionally, you'll run into a slate that slides into place and has no exposed side in which to nail. Nail it with the hidden nail in the slot technique (covered with a piece of bib flashing) that is used for basic repairs, or use a slate hook (page 223). Cut replacement slates as you need them with your slate cutter, and use matching slates. Make your cuts right there in the valley; don't go down onto the ground to cut the slates.

Be very careful when using soft copper valley flashing to not drop any sharp object on the valley metal. Soft copper punctures easily and a dropped slate hammer or ripper (or slate) can easily poke a hole in it. Furthermore, be careful when working over a soft copper valley. If you sit or stand on it and put too much weight on it in the wrong way, you can permanently buckle it, and you don't want to do that. Stainless steel valleys and heavy aluminum (.040") are much stronger than soft copper and can take a lot more punishment. Half-hard copper is preferable to soft copper in valleys, and one of the best valley flashings available is 20 ounce half-hard copper (although 16 ounce is good too).

TOP FLASHING ON A SHED ROOF DORMER

Figure 14.22

▼ **BEFORE** - The entire length of the old copper top flashing is pitted, leaking, face-nailed, and spot-tarred.

▼ **AFTER** - New 20 ounce copper top flashing on this dormer will last for generations.

▼ 1 - Number 3 rows of slates 1A, 1B, 1C, 2A, 2B, 2C etc - (number random width roofs only), remove top row with slate ripper.

▼ 2 - Carefully remove lower slates by pulling out the nails with a flat pry bar (continued on next page).

▲ 3 - Work from one end toward the other when removing the slates. Starter slates are underneath 1st row and here have no shim.

▲ 4 - Expose and remove the old flashing. Work off planks if the roof is not too steep, otherwise use roof jacks and planks.

▲ 5 - Remove any bad slates that were under the flashing.

▲ 6 - Replace the bad slates and sweep the roof.

▲ 7 - Nail in the new flashing (in this case 20 ounce copper bent over the edge of a ladder) Underbend flashing so it will fit tightly.

▲ 8 - Begin re-slating in the reverse order, put same slates in original spots. Align with weather marks. Avoid nailing through metal.

▲ 9 - This flashing is bent 6" on dormer and 10" on main roof, and is nailed along top edge only.

▲ 10 - Top row of slates are fastened with nail-in-slot and bib flashing technique (Page 223).

Figure 14.23
A skylight that extends above the roof surface is flashed according to the same principle as a chimney discussed earlier in this chapter.

Flat roofed (shed roofed) dormers on slate roofs can sometimes cause problems because they don't have the slope that the rest of the roof does and therefore rain and snow don't drain off of them fast enough. This becomes a real problem when the flashing that ties the shed roof to the main roof becomes deteriorated and begins to leak. Once again, however, this flashing can routinely be replaced, and this is the beauty of a slate roof - it can be taken apart and put back together again whenever necessary.

The procedure for replacing the top flashing on a shed roof dormer is basically the same as for replacing any flashing: remove the slates to completely expose the flashing, pry the old flashing off the roof, replace it with new metal, then replace the slates. When replacing the slates, substitute good ones for any that are cracked, broken, or tarred.

Again, use the same tricks: number the slates with a nail when random width slates are used (numbering isn't necessary when uniform slates are used), keep the slates nearby after removing them so they'll be handy when you replace them, use the same slates as much as possible when re-slating, and put them back in the same place when they're random widths. Figure 14.22 shows a top flashing replacement sequence on a random width roof.

SKYLIGHTS

Skylights fall into two general categories as far as flashing is concerned: those that protrude above the surface of the roof, and those that do not. The ones that protrude above the roof should be flashed in a manner similar to a chimney as described earlier in this chapter - the sides should be step flashed either with one-piece step flashing or step and counter flashings; the bottom should have an apron flashing; the top should have the largest piece of flashing which extends well up under the slate (at least 12"); and all four corners should be folded to prevent leaking (Figure 14.23). Some

Figure 14.24
Skylights can be laid flush with the roof, as if the glass were a big shingle. Flashing is required only on the bottom. Interior panes of glass are advisable on this type of skylight to avoid condensation. See Figure 14.25, next page.

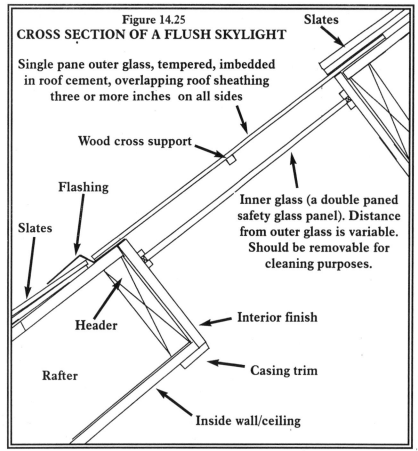

Figure 14.25
CROSS SECTION OF A FLUSH SKYLIGHT

Single pane outer glass, tempered, imbedded
in roof cement, overlapping roof sheathing
three or more inches on all sides

Slates

Wood cross support

Flashing

Slates

Inner glass (a double paned
safety glass panel). Distance
from outer glass is variable.
Should be removable for
cleaning purposes.

Header

Interior finish

Rafter

Casing trim

Inside wall/ceiling

of the standard commercial sky-lights come with pre-formed flash-ings which may need to be modi-fied to fit your particular situa-tion. Flashings are easy to make, so one may be better off making their own rather than buying expensive, generically pre-formed skylight flashing. Remember this rule when making step flashings: *the bottom of the step flashing should line up with the bottom of the slate above it, and the top of the step flash-ing should line up with the top of the slate below it.* If the step flashing must be fastened (nailed) to the roof, then allow an inch or two of flashing to extend above the top of the slate below it for nailing. If it fastens to the skylight, that won't be necessary.

Skylights that lay flush with the roof usually fall into the "home-made" category, which means that they're relatively inexpensive, and if done right, work well. The outer piece of glass on the skylight is laid into the roof as if it were a big shingle. It's flashed only on the bottom edge, while the sides and top are overlapped by slate. This piece of glass should be tempered safety glass, and should overlap the roof sheathing by three inches minimum on all four sides. It is laid directly on the felted roof sheathing on a *generous* bed of roof cement, and held in place by the heads of roofing nails tacked in around it on the corners. Wider pieces of glass may require a wooden support or two (a horizontal 1x1 for example) to prevent collapse under unusual snow weight, or when a cat's up there walking on it (Figure 14.25). It's important to know that a single

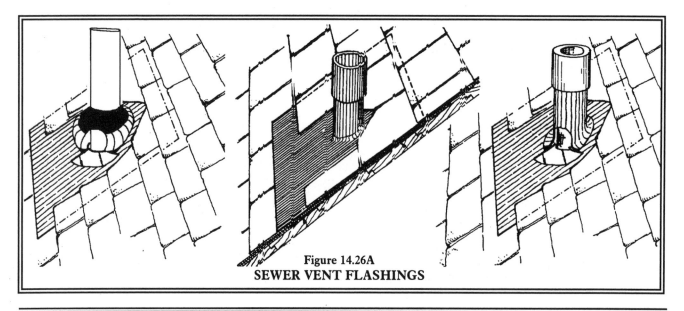

Figure 14.26A
SEWER VENT FLASHINGS

piece of glass, when used as a skylight, will condense moisture on its interior surface, and drip water when used on a heated building in cold weather. The solution to this is simply to install another piece of glass interior to the sky-light, and a double paned glass panel is even better (as far as heat retention is concerned). The drawback to "home-made" flush skylights is that they don't open and can't be used for ventilation, and may be difficult to clean. However, they will greatly increase the amount of natural light entering a room and therefore reduce the amount of electricity needed for lighting.

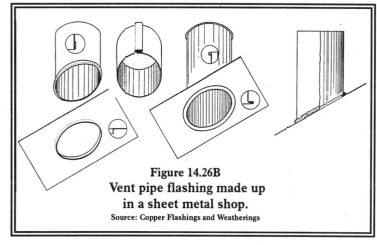

Figure 14.26B
Vent pipe flashing made up
in a sheet metal shop.
Source: Copper Flashings and Weatherings

SEWER VENT PIPES

Nearly every home and building has at least one sewer vent pipe protruding through the roof. The flashing on these pipes wears out eventually and must be replaced. Most of the old flashings were made of either copper or lead, lead being the more common. Today they can be replaced either with pre-fabricated lead, pre-fabricated aluminum/neoprene, rubber, or made-to-order lead or copper flashings. The most durable are the lead and copper flashings, although lead carries with it the hazard of handling a toxic metal. The most widely available flashings are either aluminum/neoprene or rubber, because these are typically used on asphalt shingle roofs and can be bought at most building supply centers. Both lead and aluminum flashings are available through *Oatey Co., 4700 West 160th Street, P. O. Box 35906, Cleveland, Ohio 44135 (800) 321-9532.* Lead and/or copper vent pipe flashings may be available through your local plumbing contractor, building supply outlet, or roofing contractor.

Vent pipes come in a variety of diameters, the more typical ranging between 1" and 4", therefore it's necessary to measure the pipe before ordering a pre-fabricated flashing. If no suitable flashing is available, try having one made at a local sheet metal shop.

When replacing old vent pipe flashings, the same flashing rules apply: remove enough slates from the roof to completely expose the old flashing, remove the old flashing and replace it with new, then reslate around it making sure that the slates and flashing overlap in the direction of water flow. On steep roofs the flashing may have to be forcibly bent to align properly with the slope of the roof.

STEP FLASHINGS

Step flashings go bad on dormer sides on old slate roofs after a century or so and must be replaced. Usually the sides of the dormers are covered with slate as well as the roof, so replacing the step flashings is simply a matter of removing the slates from the sides of the dormers (after numbering them), then, starting at the top, removing enough slates from the roof to completely expose the pieces of step flashing, one at a time. The step flashing is then pried off one at a time, until all are removed, then they're replaced one at a time as the slates are nailed back in place. After all the step flashings are replaced, the slates on the side of the dormer are nailed back on - an easy job if they were numbered ahead of time (e.g. row A1, A2, A3, B1, B2 etc). The flashing can be nailed either to the side of the dormer or to the roof when being installed.

When installing step flashings, remember the simple rule: line up the bottom of the step flashing with the bottom of the slate above it, and the top of the step flashing with the top of the slate below it. Extend the top of the step flashing up another inch or two for nailing purposes if desired. Lay the step flashing about 3"-4" horizontally on the roof and about 2"-4" vertically up whatever it is you're flashing.

Some dormers don't have slate on the side and there is no way to get the step flashing underneath the siding. In that case you may have to be content with tightly nailing the step flashing to the side of the dormer against a backing of silicon caulk or roof cement, leaving the vertical section of the step flashing exposed, or covered by a piece of wood trim. This is a last resort, however, as both vertical and horizontal sections of the step flashing should be covered by slate and siding. Counter flashing can also be used to cover the vertical sections of step flashings, as discussed earlier in this chapter, and the counter flashing can be bent and set into a groove cut in the siding.

Figure 14.27
STEP FLASHING ON DORMER SIDE

SOLDERING

Soldered flashings are not required when doing general slate roof restoration. However, on some occasions in unusual circumstances one may wish to solder something (e.g. a finial, a built in gutter, a cupola, or a home-made vent pipe flashing) so a brief review of soldering principles is appropriate.

Simply stated, soldering is the act of joining two pieces of metal with a molten metal or metal alloy. For our purposes, soldering involves joining copper to copper, or lead-coated copper to lead coated copper, terne to terne, terne coated stainless to TCS, or stainless to stainless, using lead/tin solder. There are enough problems involved in soldering aluminum that the Architectural Sheet Metal Manual simply states that it cannot be soldered.

The official definition of soldering is *"a group of welding processes which produces coalescence of materials by heating them to a suitable temperature and by using a filler metal having a liquidus not exceeding 450 degrees C. (840 degrees F.) and below the solidus of the base materials. The filler metal is distributed between the closely fitted surfaces by capillary attraction* [American Welding Society Soldering Manual, 1978, p. ix]." The solidus temperature is the highest temperature at which a metal or solder is completely solid, while the liquidus temperature is the lowest temperature at which a metal or solder is completely liquid.

If your mind is still functioning after reading that definition, then let's add another. "Wetting" is when a molten solder leaves a continuous, permanent film on the base metal surface, sometimes referred to as "tinning." Wetting makes the soldering action possible, and although pure lead doesn't readily wet (or adhere to) either copper or steel, a tin/lead solder readily wets both, which is why solder is made from a tin/lead combination.

Figure 14.28 Lap seam soldering: Preformed terne coated stainless stock is trimmed to fit (top left); then set in place (top right and bottom left); riveted (bottom center); and soldered with ruby fluid flux and 50/50 solder (bottom right). The rivets are also soldered.

The solder is drawn into the joints by capillary attraction (also known as capillary action), and a space between the joints of about .15mm or .005 inches is suitable for most work. A clean surface free of oxidation is critical to ensuring a sound soldered joint, and some fine steel wool will clean most surfaces adequately, prior to soldering. Solder that is 50% tin and 50% lead is suitable for use on copper, lead sheet, galvanized steel, stainless steel, terne metal, and terne coated stainless.

Prior to soldering, a flux is applied to the surfaces for the purposes of additionally cleaning away oxides, preventing re-oxidation, and promoting wetting of the surface. Fluxes fall into three general categories: corrosive, intermediate, and non-corrosive, and the mildest flux that works should be the one chosen for the job. Corrosive fluxes remain chemically active after soldering and can cause corrosion to the joint, which is why they're not used for electrical applications (for example).

Highly active and corrosive fluxes must be used on stainless steel, high alloy steels, and galvanized steel because they have hard oxide films, although rosin (non-corrosive) fluxes are satisfactory in most cases for soldering terne metal. Fluxes such as Ruby Fluid contain zinc chloride and are considered corrosive fluxes. The soldered areas where a corrosive flux was used should be washed after soldering.

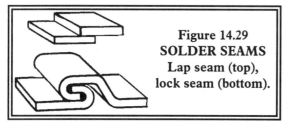

**Figure 14.29
SOLDER SEAMS**
Lap seam (top),
lock seam (bottom).

Copper tends to form less tenacious oxide films and therefore mild (non-corrosive) fluxes are suitable when soldering it. Noncorrosive fluxes all have rosin as a common ingredient.

After the flux is applied, heat must be applied to the metal to be soldered. This is typically done by heating a soldering iron and placing the soldering iron on the metal with the solder in close proximity. The solder will then flow into the cleaned, fluxed joint as the iron, and solder, are slowly moved along the joint. It may help to wet the metal (with solder) before soldering, although wetting and soldering can take place in one operation. This is a part of the operation that has a few variables, such as amount of heat, type and size of soldering iron, etc, that will vary according to type of metal being soldered, and these details must be ascertained by the artisan doing the work. For example, copper may require a hotter iron than steel because it has a greater rate of conductivity, while stainless steel may require a cooler soldering iron and longer contact. Practice soldering on a piece of scrap before doing the actual work if you need to figure out these details.

After soldering, the soldered joint should be encouraged to cool as rapidly as possible without cooling so quickly as to warp. Non-corrosive, rosin-based flux residues do not need to be removed after soldering. Zinc chloride fluxes can be removed by flushing with a 2% hydrochloric acid solution, then with hot water containing some sodium carbonate, then with clean water.

It's important to remember that all soldering fluxes give off fumes and smoke when heat is applied, and these may be toxic. Metals such as cadmium, lead, zinc and their oxides may also be toxic when present in the atmosphere as fumes or dust. Therefore, adequate ventilation is imperative when soldering any metal. Soldering fluxes containing zinc chloride may produce severe burns and dermatitis if allowed to remain on the skin for any period of time.

Two typical solder joints related to flashing are the lap seam and the lock seam (Figure 14.29). The lap seam can be joined with rivets of a compatible metal before soldering, although the joint as well as the rivets must be soldered (Figure 14.28). The flat lock seam can be cleated to an underlying structure before soldering, if needed.

[**Propane soldering pots** (for heating soldering irons) are available from *Insto-Impala Inc., 2201 W. 110th St., Cleveland, Ohio 44102, Ph: 216-251-3334*; **Propane powered soldering irons** (Red Dragon Brand) are available from *Flame Engineering, Inc., P. O. Box 577, Lacrosse, KS 67548-0577, Ph: (913) 222-2873 or (800) 255-2469, Fax: (913) 222-3619*. Or check your local roofing supply outlet.]

REFERENCES - CHAPTER 14 (FLASHING)

• Copper Flashings and Weatherings - A Practical Handbook (1951); Published by the Copper Development Association, Kendals Hall, Radlett, Herts, England.

• Slate Roofs (author unknown) 1926; available from Vermont Structural Slate Co., Fair Haven, Vermont 05743; an abbreviated version is also available from Hilltop Slate Co., Middle Granville, NY 12849.

• Architectural Sheet Metal Manual (Fifth edition, 1993); Sheet Metal and Air Conditioning Contractors National Association, Inc, 4201 Lafayette Center Drive, Chantilly, Virginia 22021.

• The NRCA Roofing and Waterproofing Manual (Third Edition - 1990); National Roofing Contractors Association, O'Hare International Center, 10255 W. Higgins Road, Suite 600, Rosemont, IL 60018-5607; Ph: 708-299-9070; Fax: 708-299-1183.

• Soldering Manual, second edition, revised (1978); American Welding Society, Inc., 2501 N. W. 7th Street, Miami, FL 33125.

Chapter Fifteen
Chimneys, Roof Accessories,
and Miscellaneous

C himneys on old slate roofs present a variety of maintenance and repair issues. First, the flashing wears out and must be replaced, which is discussed in Chapter 14. Secondly, the mortar becomes soft, especially above the roof-line, and the top of the chimney must be rebuilt before it falls apart. This is a routine maintenance job on old slate roofs, and in some cases the chimney must be rebuilt from the attic floor, rather than just from the roof-line. Third, many old chimneys no longer serve a functional purpose in the house (they're no longer used), so they should be removed to below the roof-line and the roof closed up, and this is another routine job on old slate roofs.

One old style of chimney construction often found in old farm houses involves building the chimney on a wooden platform elevated off the floor of the house. These chimneys may not even have a wooden pedestal underneath, but may be supported by a wooden shelf propping the massive brick structure five feet off the floor! Most chimneys, however, are built from the basement floor up through the house and on through the roof.

Older chimneys were typically built of 4" x 8" standard thickness brick that were solid (they had no holes in them). Therefore, when rebuilding an old chimney, the same size and type

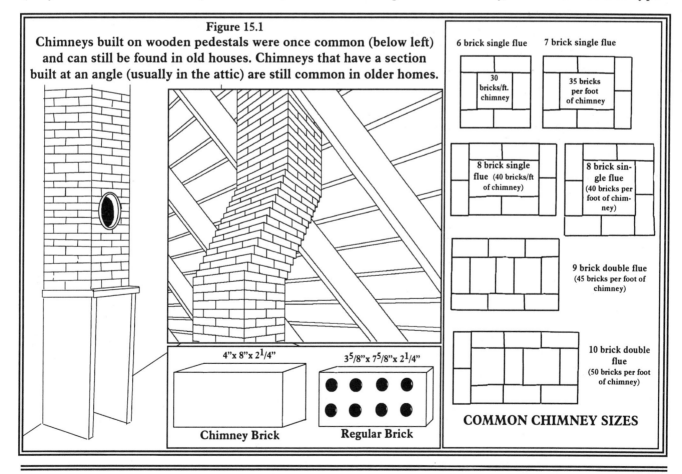

Figure 15.1
Chimneys built on wooden pedestals were once common (below left) and can still be found in old houses. Chimneys that have a section built at an angle (usually in the attic) are still common in older homes.

6 brick single flue

7 brick single flue

30 bricks/ft. chimney

35 bricks per foot of chimney

8 brick single flue (40 bricks/ft of chimney)

8 brick single flue (40 bricks per foot of chimney)

9 brick double flue (45 bricks per foot of chimney)

10 brick double flue (50 bricks per foot of chimney)

COMMON CHIMNEY SIZES

4"x 8"x 2^1/4"

Chimney Brick

3^5/8"x 7^5/8"x 2^1/4"

Regular Brick

Figure 15.2
Welsh Chimneys
(Photo by author)

of bricks are recommended. They go by the name of 4x8 (usually red) "pavers," and are a little more expensive than the bricks with holes, but they offer much greater fireproofing protection because they're solid brick.

The chimneys found in most older houses with slate roofs are almost universally devoid of ceramic chimney liners, which are additions to chimneys that became popular in more modern times when bricks began to be made with holes in them to save on material costs and cut down their weight. The liners added an additional firewall for protection, although they can be dispensed with when using full sized, solid brick, which create a four inch thick solid chimney wall. In the old days, chimneys were used to vent smoke and gasses from coal and wood burning stoves, which tended to burn hot and inefficiently. Today, most chimneys vent either the more efficient "air tight" stoves, or gas/oil furnaces, both of which burn much cooler than the old-time heaters. Cool burning woodstoves are notorious producers of creosote, which can leak through brick chimneys, ceramic liners or not.

When rebuilding an old chimney or part of it, the original style should be imitated, unless the entire chimney will be rebuilt, then any style can be used. However, the old style of using solid bricks laid up without a ceramic liner is a tried and proven style, and many thousands of these old chimneys are still in service after a century of use. Nevertheless, old chimneys can be dangerous, especially if the mortar and/or bricks become soft enough to crumble and allow flames to escape to a flammable surface during a chimney fire. Chimney fires are unlikely when the chimney vents only a gas or oil furnace because these don't create an accumulation of creosote like a wood stove does. But when using an old chimney to vent a wood stove, the chimney should be inspected closely beforehand to make sure it's safe. If it doesn't appear to be safe, it should be completely rebuilt or lined (with a metal liner, for example).

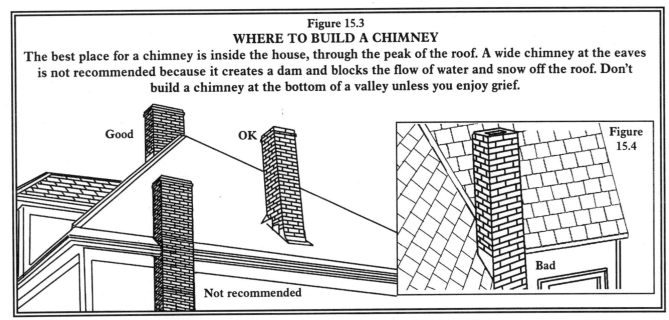

Figure 15.3
WHERE TO BUILD A CHIMNEY
The best place for a chimney is inside the house, through the peak of the roof. A wide chimney at the eaves
is not recommended because it creates a dam and blocks the flow of water and snow off the roof. Don't
build a chimney at the bottom of a valley unless you enjoy grief.

Good

OK

Figure
15.4

Not recommended

Bad

Another peculiar style of chimney building that was common in the old days involved the practice of building a section of the chimney (usually in the attic) at an angle. Once through the attic floor, the chimney was angled so the top would exit the roof directly through the center of the ridge - a style adhered to by certain ethnic builders perhaps (such as Germans, for example). It's these leaning chimneys that must often be rebuilt from the attic floor because of deteriorating bricks or mortar, which can create quite a hazard if the chimney should begin to collapse. When rebuilding a leaning chimney it should be straightened, afterwhich it will exit through the roof slightly to one side of the original hole. New roof sheathing (1" rough-sawn) is simply added to fill in the hole (after suitable framing material has been nailed into place) and the area is then reslated as the chimney is flashed. See Chapter 12 for more information on roof sheathing.

When building a chimney for a new building, be aware that there are best and worst locations to build them. The best location for a chimney is inside the structure and through the peak of the roof. Inside chimneys stay warmer, radiate more heat throughout the building, and develop less creosote problems than cooler, outdoor chimneys. The emergence of the chimney at the peak of the roof prevents any water from pooling up against the chimney. From a roofing point of view, a chimney built up against the outside of the building through the peak of the roof works just as well as one built inside the building. However, from a heating point of view the inside chimney is preferable.

An undesirable place to situate a chimney is at the eaves of the building, especially if the chimney is wide, as it creates a dam on the roof which will eventually turn into a leak (Figure 15.3). The worst place to put a chimney is at the eaves *and* at the bottom of a valley. This sort of chimney falls into the "roofer's nightmare" category, although they can be made leakproof with a carefully constructed cricket (Figure 15.4).

REBUILDING THE TOP OF A CHIMNEY

The first step in rebuilding a chimney top is setting up roof scaffolding. This can be easily and quickly done by using hook ladders and ladder jacks as illustrated in Figure 15.5. When the chimney is halfway down the roof instead of at the peak, hook ladders alone with two pairs

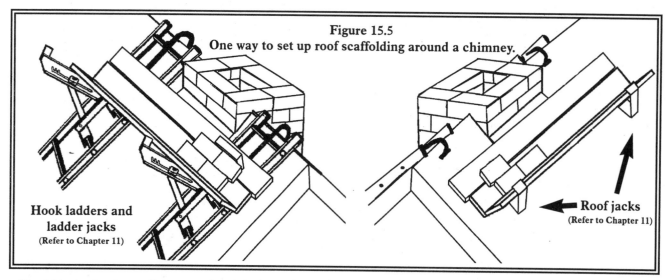

Figure 15.5
One way to set up roof scaffolding around a chimney.

Hook ladders and
ladder jacks
(Refer to Chapter 11)

Roof jacks
(Refer to Chapter 11)

of ladder jacks (one above and one below the chimney) will work. If hook ladders can't be used (because there's no ridge above the chimney), roof jacks and planks will have to do. The danger of using any roof scaffold is in overloading it (with bricks for example), and/or carelessly installing it. Double check everything before any weight goes on the scaffold, then make sure it doesn't become overloaded.

Once the roof scaffold has been set up, the old chimney top should be removed and new bricks should be used to rebuild it. Old tops can usually be taken apart by hand as the mortar becomes soft after a century of weathering, although some of the more stubborn chimneys may require a small sledge hammer and a flat pry bar to get the bricks apart. The top should be taken down to just below the roof line, far enough to remove all of the old flashing. All old mortar should be removed from the top of the last remaining course of bricks, then the bricks brushed clean of dust in preparation for new mortar.

When a chimney top is rebuilt, it makes sense to reflash it too, because it's much easier to flash a chimney when rebuilding it than it is to try to flash it after it's been built. The old flashing, then, is completely removed as are the slates surrounding the chimney covering the flashing, as well as any bad slates (broken or tarred). It's always easier to completely clean all the bad stuff out of the way before beginning to put the roof and chimney back together.

REBUILDING THE TOP OF A CHIMNEY

TOOLS YOU WILL NEED: Mortar box and hoe for mixing mortar, small mortar box for use on the roof to hold the mortar (a plastic dish pan will do); pointed masonry trowel; 5 gallon plastic buckets for carrying mortar and possibly bricks up onto the roof; brick tongs for carrying brick (optional but recommended); 2 foot and 4 foot levels; striking tool for striking mortar joints after bricks are laid; slate ripper; slate cutter; slate hammer; flat pry bar; small sledge hammer for knocking old top apart (may not be needed); bucket of clean water and rags for wiping chimney clean after bricks are laid (important); tin snips; flashing (sheet metal) tongs (optional when using 16 ounce soft copper).

EQUIPMENT: Ground ladders, hook ladders, roof jacks, scaffold planks.

MATERIALS: Regular masonry cement (mortar); mason's sand; new bricks; flashing material (16 ounce soft copper recommended); water; slates that match the roof.

MORTAR MIX RATIO: One part masonry cement to three parts masonry sand. Mix dry first, then add water slowly to consistency of buttery mud.

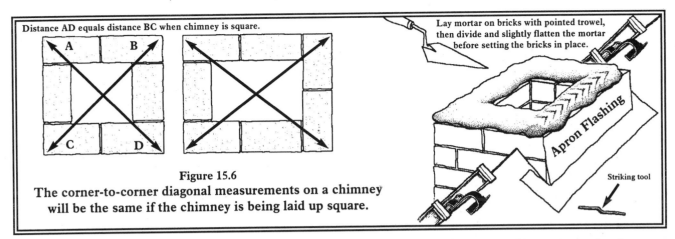

Distance AD equals distance BC when chimney is square.

Lay mortar on bricks with pointed trowel, then divide and slightly flatten the mortar before setting the bricks in place.

Apron Flashing

Striking tool

Figure 15.6
The corner-to-corner diagonal measurements on a chimney will be the same if the chimney is being laid up square.

When building a chimney indoors, such as through an attic, run strings where the four corners of the chimney will be and use them as a guide when laying the brick (Figure 15.7). This saves quite a bit of time because the sides don't need to be plumbed with a level as the chimney is built, although the courses still need to be checked for levelness as each course is laid. Furthermore, when building a chimney from the bottom up, make sure there is a clean-out door somewhere near the base of the chimney (in the basement), and try to leave a drain in the bottom of the chimney so any rain water that accumulates down there will have a place to go (heavy downpours will allow a lot of water to enter a chimney, even when capped).

When the bricks and tools are in place, the mortar is mixed. It's usually a good idea to have one person mixing mortar and hauling materials and another laying the bricks. **The mortar is mixed according to a ratio of one part masonry cement to three parts masonry sand.** These materials are always mixed together dry first until the mix is a uniform grey color, then water is slowly added until the correct consistency is obtained. If the mix is too wet the mortar will run all over the chimney and make a mess, but if it's too dry it will make the brick laying job more difficult. If it is too dry, however, a little water can be added by the brick layer to get the mortar to the right consistency, which should be like thick pudding.

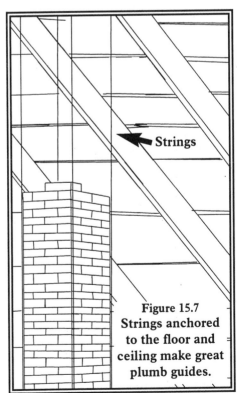

→ Strings

Figure 15.7
Strings anchored to the floor and ceiling make great plumb guides.

The mortar is then laid on the clean brick surface with the pointed trowel, then cut into two halves and slightly flattened with the point of the trowel, and the bricks are laid in place. After the first brick is laid, the next must be buttered on the end that will abut the first brick, before setting it into place. Every course of bricks must be staggered in relation to the one below it so the joints don't align. The courses must be checked with a level as they're laid, for levelness and for plumbness. The two foot level can be used to plumb the first two feet of the chimney, then a four foot level is necessary. The chimney should also be checked for squareness, either by using a framing square, or by measuring the diagonals (diagonal corner to diagonal corner), which will measure exactly the same if the chimney is square. Smooth the excess mortar on the inside of the chimney against the bricks with the trowel.

Fold the flashing into the mortar joint a distance of about two inches. Start with the apron flashing, then the lower corners (which are folded around the corners of the chimney - see page 246), then the sides. Sixteen ounce soft copper is recommended as flashing material because it's long lasting and easy to work with (it can be bent by hand). You'll have to set the apron flashing in place, then lay another course of bricks (maybe two), install the corner flashings, then lay a course or two of bricks, etc. This is a tedious job that benefits greatly from experience.

After the bricks have been laid, the mortar joints must be struck with a striking tool (page 272, #7). The joints must be struck before the mortar gets too hard, so it's usually a good idea to strike the joints as you go along, allowing them to set up somewhat first. As soon as the job is completed (most single or double flue chimney tops can be reflashed and relaid in a day by two workers), the bricks should be carefully washed with clean water and a clean rag. There is no easy way to do this because if the mortar gets wet, it will run over the face of the bricks and make a mess, so the wet rag must be wrung out and carefully wiped across the face of *each* brick to remove any traces of mortar. Otherwise, when the mortar dries, it will appear as white splotches or streaks on the bricks, and no matter how well the bricks were laid, a messy chimney forever looks bad. Once the mortar does dry out it's almost impossible to clean it off without using an acid solution, which is not advisable on a slate roof due to the effect of the acid on the flashing. Take

Figure 15.8 DRIPSTONES
Stone shelves built into the chimney divert water away from the bottom of the chimney, acting much the same as flashing. Dripstones are a common sight in Wales where both chimney and building are constructed of the same material.
(Photo by author)

a little extra time to carefully clean *each* brick when the job is done, before the mortar dries, and the chimney will be beautiful.

Finally, after the chimney top rebuilding job is done, you must go down into the basement of the house and find the clean out door at the bottom of the chimney (and you better hope there is one - otherwise you may have to make one). There you will need to remove any mortar or other debris that has fallen down the chimney when the top was rebuilt. Otherwise you risk plugging the chimney, and a plugged chimney is dangerous because it can allow carbon monoxide, an odorless and lethal gas, to accumulate in the house. If you have any doubts about whether your chimney is venting properly, install a carbon monoxide detector in your home, available at most hardware stores.

One last thing - when rebuilding a chimney top, take an extra hour or so and do a fancy job. The extra time is well worth it when one realizes that the new chimney top will be visible for generations.

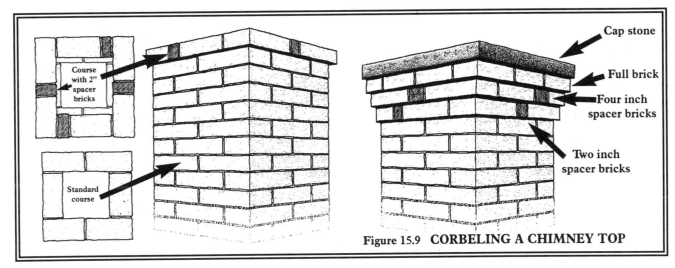

Figure 15.9 CORBELING A CHIMNEY TOP

BUILDING A FANCY TOP

Chimneys were often traditionally flared at the top in order to accommodate a chimney cap stone, which was usually a single piece of sandstone laid over the top course of bricks with a square hole cut in the middle to allow the chimney to exhaust. Bricks can be easily stepped out, flared, or *corbeled*, by simply adding a spacer brick in the course to be corbeled (Figure 15.9). A one inch spacer brick on every side of the chimney course will space that course out 1/2 inch on all four sides. A two inch spacer will space the course out an inch, etc. By successively increasing the size of the spacer brick by two inches, each subsequent course of bricks will be stepped out an inch further than the one below it. A course can be stepped out, then stepped back in, then stepped out again as shown in figure 15.10. There are an endless number of variations related to chimney top designs, and the design one settles upon for his/her own work is only limited by one's creativity. A good chimney builder will build a "signature" top on his/her chimneys, such as the one described in Figures 15.10 and 15.11.

Figure 15.10

This old style fancy chimney top is not difficult to build and the procedure is illustrated in Figure 15.12. At left is a double flue chimney on a historic house in Harrisville, PA with an asbestos roof. The chimney was a leaning chimney and was rebuilt plumb from the attic floor. At right is a 6 brick single flue chimney built new through a recycled sea green slate roof (shown on page 195). Both chimneys are flashed with 16 ounce soft copper.
(Photos by author)

1. Come up about 10 courses of brick from the top of the ridge, then set a course out an inch, then two regular, then another out an inch.

2. Set full bricks in a bed of mortar diagonally on the corners, flush on the ends. Level and straighten them.

3. Fit in the remaining diagonal bricks, lining up their outside edges and making sure they're level. Some will need to be cut short -- use a brick hammer to cut them, or use a masonry or diamond wheel on a grinding tool or electric saw.

4. Lay all the diagonal bricks in place, corners flush with the top square course, outside edges lined up, and everything level.

5. Lay a square course on top, outside edge flush with the tips of the diagonal bricks.

6. Lay a second square course on top of the last one, then lay a bed of mortar on top of it as a final protective layer. Level it and dress it up.

7. Strike the joints, then carefully clean the chimney, brick by brick, with clean water and a rag

Figure 15.11
BUILDING A FANCY CHIMNEY TOP
(See Figure 15.10)
The trick to making a decent chimney top is to take your time, and level and plumb everything. Use plenty of mortar and make sure it isn't too dry - you don't want it drying out and forcing you to rush. Slightly wet mortar makes laying small pieces easier too. A top like this will only add an hour or two (if you're really slow) to the job of building the chimney top, and since the top might last a hundred years it's worth the extra time!

REMOVING AN OLD CHIMNEY

Old chimneys lose their usefulness after a while and are best removed from the roof. Once gone, they pose no threat of leaking and require no expense or upkeep. They can be rebuilt when needed if they're only removed to just below the roofline. Such a removal job is routine and can be done so that the roof appears to never have had a chimney at all.

The job is simple and can be briefly described. As always, most of the job is in setting up a safe working platform. Again, roof jacks and planks along with hook ladders, will be sufficient in most cases to get the job done. Once you've made your way to the chimney top, you may find that the bricks come apart in your hand, which is common on old chimneys. If the chimney is stubborn, a small sledge will convince it that its days have come to an end. After the bricks are knocked apart and the chimney is lowered enough to clear the rafters, 2x4's are scabbed on to the rafters to create a place to nail the sheathing boards that will be patched in to cover the chimney hole. The sheathing consists of full 1" rough sawn lumber, and green lumber is fine.

Sometimes the rafter is set too far back in from the chimney hole to be able to scab it out

Figure 15.12   REMOVING AN OLD CHIMNEY

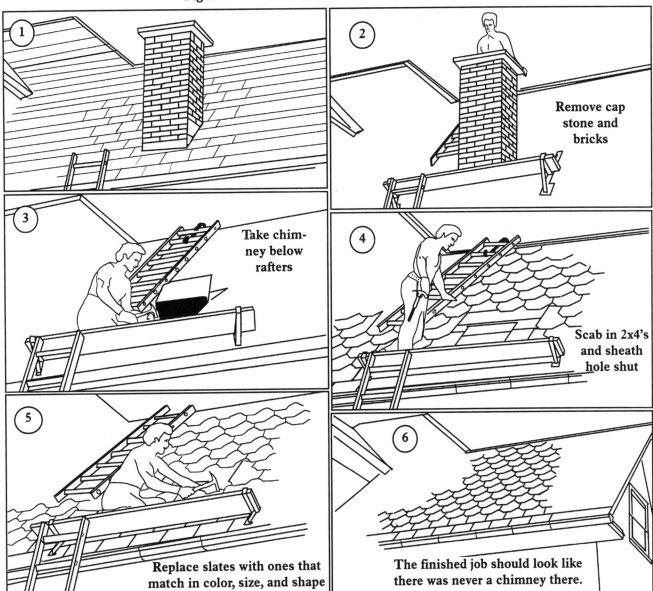

1

2 Remove cap stone and bricks

3 Take chimney below rafters

4 Scab in 2x4's and sheath hole shut

5 Replace slates with ones that match in color, size, and shape

6 The finished job should look like there was never a chimney there.

enough to nail sheathing on to it. On this rare occasion, one will have to take a saw up on the roof and cut the old sheathing back to the rafter center, then add the new sheathing nailed directly to the old rafter.

When the chimney is being taken down through the roof, the old flashing will need to be removed, as will the slates around the chimney, especially any that are tarred or broken. When the chimney has been taken out of the way and the mess on the roof cleaned up, there will be only a hole and good, full-sized, solid slates around the spot where the original chimney used to be. Once the hole is closed up with the new sheathing, new slates are nailed in to replace those missing, and the job is done (Figure 15.12).

ROOF ACCESSORIES

SNOW GUARDS

Snow guards are small, angled metal brackets permanently attached to the roof to prevent ice and snow from cascading off the roof in an avalanche. They fall into two primary categories:

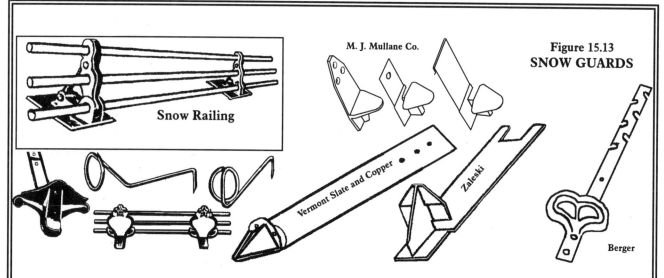

Figure 15.13
SNOW GUARDS

SOURCES OF SNOW GUARDS

Berger Building Products - 805 Pennsylvania Boulevard, Feasterville, PA 19053; Ph: 1-800-523-8852 or (215) 355-1200; Fax: (215) 355-7738. They have the handy hook-on kind that can easily be installed on an existing slate roof (in copper, stainless steel, or hot-dipped galvanized). They also have a variety of other styles and types.

M. J. Mullane Company - 17 Mason Street (P. O. Box 108), Hudson, MA 01749; Ph: (508) 568-0597, Fax: (508) 568-9227. The have a wide variety of snow guards, including bronze and aluminum cast.

Vermont Slate and Copper Services - 55 C Gonyeau Road, Milton, VT 05468; Ph: (802) 893-7703, Fax: (802) 893-1143. They have custom-made copper snow guards, leader boxes, and finials.

Zaleski - 11 Alsen Street, New Britain, CT 06053; Ph: (203) 225-1614, Fax: (203) 225-1060. Their slate snow guards protrude between the slates and hook on the top of the slate without the need for nailing.

[Also check your local roofing supply outlets.]

those that are nailed to the roof, and those that hook either on the slates themselves, or on slate nails. The ones that are nailed on must be installed when the roof is installed. Otherwise, slates must be removed from the roof, the snow guards attached, then the slates replaced. On the other hand, the kind that hook on an existing slate or nail can be attached to the roof at any time with much less effort. Some simply have a long slotted end which is slid up under the slate as if it were a slate ripper, and hooked on a slate nail. Others are slid between the slightly lifted slates until they hook over the top. The snow guard, once hooked, is left in place.

Snow guards are positioned above doorways, walkways, porches, sidewalks and anywhere where falling ice or snow may present a hazard. They're made of either galvanized steel, painted steel, stainless steel, copper, aluminum or bronze, and may be lead coated. Some come as "snow railings," an elongated version of the snow guard (Figure 15.13).

LIGHTNING RODS
(plus Weather Vanes and Finials)

Lightning rods are common on old slate roofs. They're simply nailed onto the roof right

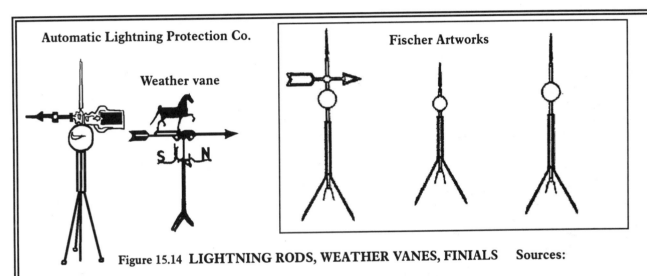

Figure 15.14 LIGHTNING RODS, WEATHER VANES, FINIALS   Sources:

Automatic Lightning Protection - 11072 W. Mohawk Lane, Sun City, AZ 85373; Ph: 1-800-532-0990; Over 200 styles of lightning rods and weather vanes.

Colonial Works (weather vanes) - PO Box 46457, Hollywood, CA 90046; Ph: (213) 460-6838.

Denninger Weather Vanes and Finials - 77 W. Whipple Road, Middletown, New York 10940-9801; Ph/fax: (914) 343-2229.

Fischer Artworks - 6530 S. Windermere St., Littleton, CO 80120; Ph: 303-798-4841/800-441-6067, Fax:795-8805; Copper and cast bronze lightning rods with copper or glass globes.

Independent Protection Company., Inc., - PO Box 537, Goshen, IN 46527; Ph: (219) 533-4116, Fax: (219) 534-3719. A variety of lightning rods, including old style (antique).

Lehman's Hardware - One Lehman Circle, Kidron, Ohio 44636; Ph: (330) 857-5757, Fax: (330) 857-5785; Email: GetLehmans@aol.com; web site: http://www.lehmans.com; A variety of both lightning rods and weathervanes.

Vanes and Things (dog motif weather vanes) - 1112 East C. Street, Commerce, OK 74339; Ph: (918) 675-4262.

Vermont Slate and Copper Services - 55 C Gonyeau Road, Milton, VT 05468; Ph: (802) 893-7703, Fax: (802) 893-1143. Finials.

[Also check your local hardware and roofing supply outlets.]

through the slate, usually with rather large (i.e. 16 penny) nails, and the nail heads are caulked or cemented to prevent leakage. The feet of old lightning rods will sometimes cause leaks because the cement over the nails has worn away, and a fresh application of roof cement or silicon caulk will cure the problem. Lightning rods are designed to divert electric current from lightning away from a building by channeling the current through a heavy copper or aluminum cable into the ground. Therefore, a lightning rod must be properly grounded or it won't work at all. Many older homes with old lightning rod systems are having the rods and cables removed. However, for those who feel that lightning rods are a necessary safety element to their roof, sources of lightning rods are listed in Figure 15.14 (page 275).

MISCELLANEOUS

SOLDERED-SEAM METAL ROOFS

Many old slate roofs have soldered seam (metal) roofs abutting them. They're low slope roofs, almost flat, and usually found on porches, bay windows, and two-story additions. These roofs are made of terne metal soldered at flat seams, hence the name "soldered-seam roofs," and they're excellent and long-lasting roofs *if they're kept painted* - most aren't. These roofs differ from "standing seam" roofs, which have an obvious metal seam that sticks up into the air an inch or two, running the length of each metal roofing panel. It's the *soldered seam* roofs that are most commonly associated with old slate roofs.

The reason they're mentioned here is because they're a maintenance problem for many owners of old slate roofs since the metal is often run up under the slates where the flat roof abuts the slate roof. If you have a slate roof with a soldered-seam roof associated with it, there are two things you should know. First, soldered-seam roofs should be kept painted with tinner's red oxide paint (or any good, exterior metal paint). If not, they'll rust and start leaking, usually at the seams. Second, if the soldered-seam roof is already leaking because the roof has long been neglected and now you have to do something about it, there is an easy and effective remedy: *liquid asphalt emulsion and fiberglass membrane.*

Liquid asphalt emulsion is an old-fashioned roofing material that works well to preserve old metal roofs when paint won't do. It's a water-based (emulsified) asphalt that must be applied when no rain is imminent as it will wash off if it is subjected to a downpour before it sets up. But once it dries it's insoluble in just about everything, and you'll soon be a believer when you try to wash dry asphalt emulsion off your hands or clothes with water, gasoline, or anything else after applying it to a roof. Liquid asphalt emulsion comes in five gallon buckets and is applied with a long-handled roofing brush. It's liberally brushed on to a clean, dry surface (like painted or rusted metal), then a layer of fiberglass membrane (which is made to be used with the emulsion) is carefully rolled out over the wet asphalt and brushed in, then a second liberal coat of asphalt is applied over top of the fiberglass. The fiberglass comes in rolls of various widths, and the three foot wide roll is most practical for larger surfaces. The whole thing needs to set up overnight, and maybe longer if the weather is too cool or too humid, then a third, final coat of asphalt is applied right over top of the whole works.

When all dry and done, the metal has a thin coating of black asphalt (which can be painted to change the color, if desired) that remains hard even in the hottest weather, and can be walked on liberally. When the asphalt wears thin a number of years down the road, brush on another coat and throw in some fiberglass over any weak areas. Be very careful to not apply asphalt over any slates - brush it up *under* the slates when you're coating a metal roof that abuts a slate roof, and make sure the fiberglass is worked up under the slate too.

Liquid asphalt emulsion will seal up your soldered-seam roof so it won't leak, and at a reasonable cost. An alternative covering for old soldered-seam roofs that is gaining popularity is rubber roofing. The problem with rubber roofing is that the rubber must also be worked up under the slates, and most rubber roof guys just slap it over top of the slate and glue it down. This creates big problems down the road when the glue lets loose - and it will. Then try to find your rubber roof guy (good luck). Properly done, a rubber roof abutting a slate roof requires that the slates are lifted, the rubber installed underneath, then the slates are relaid. Liquid asphalt emulsion is cheaper, easier to work with, available at most roofing supply outlets, and do-able for most do-it-yourselfers. But watch out - it's messy!

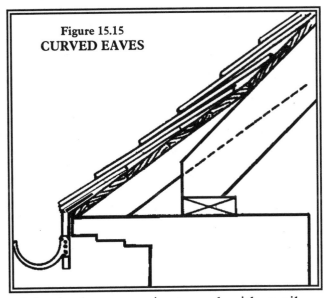

Figure 15.15
CURVED EAVES

CURVED EAVES

Curved eaves aren't very common, but you do run into them on old slate roofs occasionally. When damaged, they should be restored to their original condition by simply replacing any missing, tarred, or broken slates. The slates on curved eaves will tend to be shorter than the ones on the main roof, and may be considerably shorter in order to conform to the curve on the roof. As always, the slates must have sufficient headlap (page 201), and if so will shed water like the back of a duck.

ASBESTOS ROOFS

Asbestos roofs are mentioned here because people who have this type of roof on their buildings often mistake them for slate. Some even call them "asbestos slate" roofs. Asbestos roofing is a human-made material consisting of asbestos fibers and a binding agent of some sort. It does not have the characteristics of natural slate roofing, although it can resemble slate in appearance, especially to the layperson. Like slate roofs, asbestos roofs tend to be old, and it can be difficult to find someone to repair them. One complication associated with asbestos roofs is the fact that asbestos is now considered a toxic material because it can cause lung cancer if inhaled, and therefore the removal of asbestos roofing is expected to be done by

Figure 15.16
ASBESTOS ROOFING
Most of the old asbestos roofing is laid up in the diamond pattern shown below. A one inch square piece of flat copper with a pin sits under two overlying shingles and the pin feeds through a hole in the bottom of the top shingle and bends over to secure it. The top of the shingle is nailed in two places.

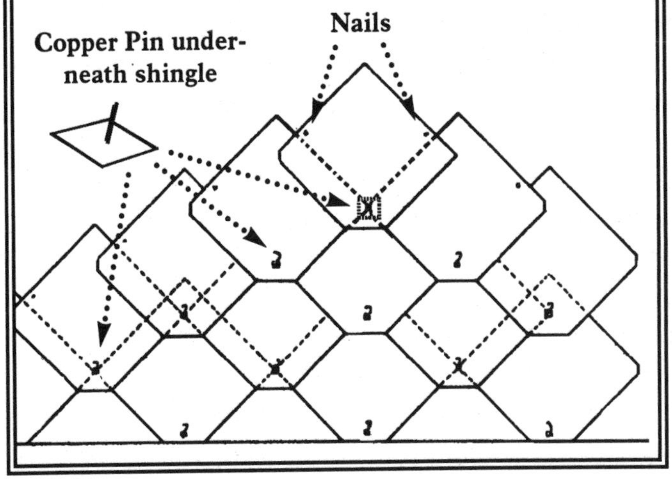

Copper Pin underneath shingle

Nails

toxic waste professionals. This puts owners of asbestos roofs in a bind - they must either keep their asbestos roofs, pay big bucks to have the roof removed by toxic waste professionals in moon suits, or illegally remove the roofing themselves and get it to a landfill unnoticed.

The bright side of this dilemma is that asbestos roofs can be maintained for quite some time *as long as replacement shingles are available* to replace any that become broken over time. Asbestos roof owners who want to keep their asbestos roofs should acquire a collection of asbestos shingles, and since you can't buy them new you may have to advertise for them (i.e. "wanted - used asbestos roof shingles"). If a stash of asbestos shingles is available, then asbestos roofs can be maintained much the same as slate roofs are maintained -- with hook ladders to work on the roof, slate rippers to remove the old tiles, and slate cutters to cut them.

Asbestos roofing cannot be punched with a slate hammer like a slate can because it's more brittle and tends to crack, but it can be drilled, or punched with a slate cutter that has a hole punch on it. Asbestos shingles can't be replaced like slates can with a hidden nail in the overlying slot covered by flashing, *unless* the asbestos shingle is rectangular and laid like slate. Mostly, however, asbestos shingles are laid in diamond patterns and have no overlying slot, so replacing them is a bit more problematic and usually requires a face-nail, and some roofing cement or silicon to hold the shingle in place.

The original asbestos shingles are nailed with two nails, and the diamond patterned ones also have a copper clip that hooks through the bottom of the shingle to keep the wind from getting underneath and lifting it (Figure 15.16). Asbestos shingles are brittle enough that they can develop hair-line cracks not visible from the ground, and these cracks will leak. The roof must be looked over very closely to find these cracked shingles. When replacing the shingle, the old, broken one is removed with the aid of a slate ripper, which pulls out the nails, then a new one is slid into place over a bed of roof cement or silicon, and nailed through the hole at the bottom (where the pin on the copper clip had come through) with an eight-penny nail, which is caulked over with clear silicon. It may be beneficial to drill through the underlying shingles before driving the nail through, as the nail may crack the brittle shingles. A simple non-electric hand drill will do.

CERAMIC TILE ROOFS

Ceramic tile roofs deserve an entire book of their own. They're briefly mentioned here because slate roofers are also called on to repair old tile roofs (and asbestos too, by the way), and a lead for information and materials can save someone a lot of research time: *Sources of new ceramic tile and accessories: Ludowici-Celadon Inc., 4757 Tile Plant Road, PO Box 69, New Lexington, Ohio 43764; Phone: (614) 342-1995.* Used tiles are available from *Renaissance Roofing* (page 218).

BUILT IN GUTTERS

Built-in or box gutters are very common on buildings that have old slate roofs, especially on the porches. The gutters, like soldered-seam roofs, were typically made from soldered terne metal. They rust and corrode when not kept properly cleaned and painted, and many of these old gutters leak. They can be replaced, however, by carefully measuring the wooden frame holding the gutter, then having the metal stock made-up at a local sheet metal shop in ten or twelve foot lengths. Use terne-coated stainless or some other metal that won't corrode over the years, overlap the metal sections by an inch and a half or two inches, rivet the pieces together with stainless steel or copper rivets, then solder the joints *and* the rivets. The procedure is illustrated in Figure 15.17, and some pertinent photos are shown in Chapter 14, page 263.

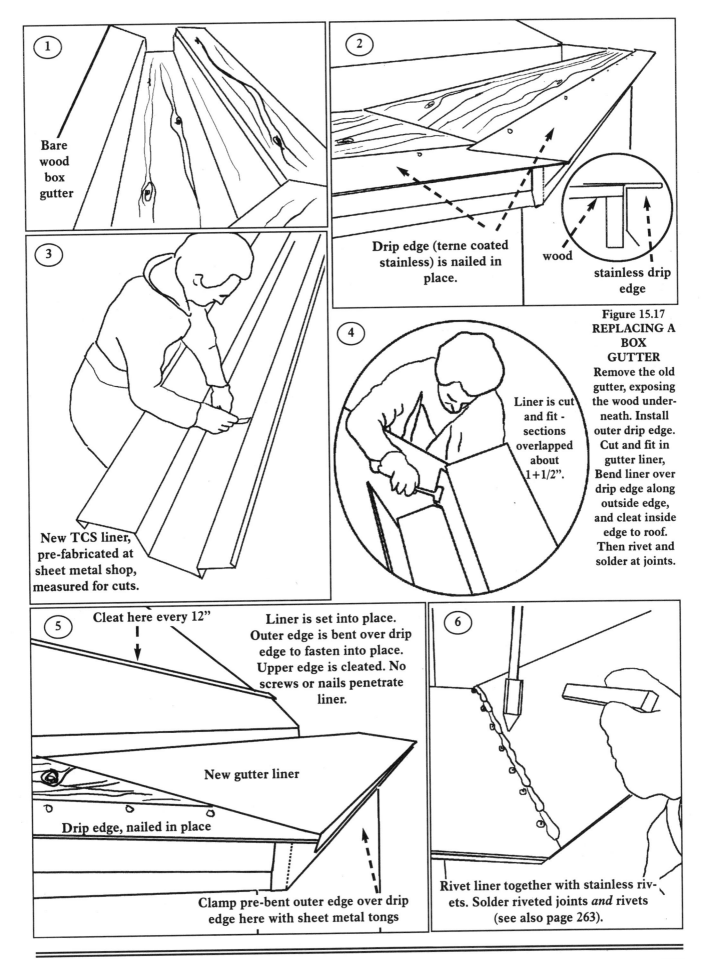

1 Bare wood box gutter

2 Drip edge (terne coated stainless) is nailed in place.

wood

stainless drip edge

Figure 15.17
REPLACING A BOX GUTTER
Remove the old gutter, exposing the wood underneath. Install outer drip edge. Cut and fit in gutter liner, Bend liner over drip edge along outside edge, and cleat inside edge to roof. Then rivet and solder at joints.

3 New TCS liner, pre-fabricated at sheet metal shop, measured for cuts.

4 Liner is cut and fit - sections overlapped about 1+1/2".

5 Cleat here every 12"

Liner is set into place. Outer edge is bent over drip edge to fasten into place. Upper edge is cleated. No screws or nails penetrate liner.

New gutter liner

Drip edge, nailed in place

Clamp pre-bent outer edge over drip edge here with sheet metal tongs

6 Rivet liner together with stainless rivets. Solder riveted joints *and* rivets (see also page 263).

Index

A

Acme Run - 27, 102
Acro Building Systems - 179, 180, 185
aerial carrier - 90, 91
aesthetics - 13
AJC Hatchet Co. - 169
Akemi Adhesives - 217
Albion Run - 27, 102
    gray bed - 27
    veins in - 102-103
Alcoa - 220
Algiers - 129
Alonson Allen (Col.) - 77, 79, 80
aluminum - 182, 204, 240, **241**, 243
    sewer vent flashings - 261
    soldering - 262
    sources of - 237
    spouting - 220
Amari Metals - 237
American slates
    Cutting into Welsh market - 59
anodic - 242, 243
Appalachian Range - 132
apron (chimney) - 247
Architectural Sheet Metal Manual - 244, 262
Arkansas - 26, 43
    Mena - 43
    Montgomery Co. - 43
    Norman - 43
    Polk Co. - 43
    Slatington - 43
Arlington National Cemetery - 142
Arvonia (VA) - see "Virginia"
asbestos roofs - **277-278**
asbestos tile - 59, 62
asphalt shingles - 9, 11
ASTM - 30
Australia - 25, 143
Austria - 25
Automatic Lightning Protection Co. - 275

B

badrockmen - 63
Bali - 25
Bangor (PA) - see "Pennsylvania"
Bangor slate - 104, 152
Bangor, Wales (see "Wales")
bargain - 67
barns - 187, 225
barn roofs - 165, 189, 225
Bartow Co. (GA) - 43
basic slate repair - 223-223
bats - 153, 157, **161**, 234
bedding plane - 22
beds - 102
bees - 157, 234
    bumble bees - 158, 160
    carpenter bees - 158, 160
    honey bees - 158, 159
    hornets - 158
        bald-faced - 160

mud daubers - 158, 160
    paper wasps - 158, 160, 234
    yellow jackets - 158-159
Belden - 173
Belgium - 25, 59
Benjamin Obdyke Co. - 234, 237
Berger Building Products - 274
Berridge, Neil - 212
bib flashing - 182, 224, 257
bioregion - 191
black powder - 85, 92
black slate - 31, 101, 103, 107, 110, 121, 127,
    132, 142, 152
blacksmith - 82, 105
Blaenau Ffestiniog (see "Wales")
blasting powder - 63
block - 107, 108, 111
block cutter - 105
Boer War - 59
box gutters - **278-279**
    replacing - 279
brachiopods - 128
brass - 243
bricks - 265, 266, 268
    corbeling - 271
    laying - 269
brontosaurus - 24
bronze - 243
Brownsville, ME - 43
Buckingham (VA) - see "Virginia"
Buckingham-Virginia Slate Corp. - 128, 130,
    131, 218
Buhl Mansion - 145
built-in gutters - **278-279**
    replacing - 279
Burgoyne's Cove - 77
butt and fireside - 84

C

Caernarvon (see "Wales")
California - 43
    Eldorado Co. - 43
    Kelsey - 43
cadmium - 243, 264
Camara Slate Co. - 96, 185, 218
Cambrian Age - 22, 54, 135, 143
Canada - 25, 139, 143, 146
    Bedard Quarry - 143
    Brantford - 144
    Burgoyne's Cove - 143
    Danville - 143
    Kingsbury - 143
    London - 144
    Melbourne Quarry - 143
    Montreal - 144
    Newfoundland - 77, 143, 144
        map - 143
    Newfoundland Slate Co. - 144
    New Rockland Quarry - 143
    Richmond - 143
    Rimouski - 143
    Steel Quarry - 143
    St. Lawrence River - 143

Toronto - 144
    Trinity Bay - 143
    Trinity Slate - 144
    Walton Slate - 143
Caparosa - 124
capillary action - **230**, 262, 263
cap flashing - see "flashing, counter"
cap slates - 203, 219, 224, 235-236
carbon - 240
carbon monoxide - 270
Cardiff (MD) - see "Maryland"
carpenter bees - see "bees"
carriage - 89
    self-dumping - 89
Cassady-Pierce Co. - 185, 237
cast iron - 243
Cathedral Gray - 44
cathodic - 242, 243
caulk - 184
cave - 13
caveman - 154
cavepeople - 13
ceramic tile roofs - 278
chalk lines - 202, 203, 204, 205, 209, 255
Chapman (PA) - see Pennsylvania
Chapman slate - 106, **109**
    identifying - 103
    quarries - 110
Chapman Slate Co. - 110
Chapman Standard Quarry - 110
Chestnut Level Pres. Church - 117
chicken ladders - 164, 178
chimney - 157, 185, 192, **265**
    apron - 247
    brick - 265
    cleaning (outside) - 270
    clean out door - 270
    corbeling - 271
    cricket - 248, 249
    drain - 269
    fancy top - **271**
            building - **272**
    fires - 266
    flashing - 239, **244-250**, 269
    hazards - 161
    leaning - 265
    liners - 266
    on pedestal - 265
    plumbing - 269
    rebuilding the top - **267-272**
    reflashing - **249-250**
    sealing - 232-233
    sizes - 265
    top, removing - **273**
    Welsh - 265
    where to build - 267
China - 25
chips - 87
chisels - 85
chlorite - 21
chromium - 240
Civil War - 131, 134
CK slate hammers - 174, 235
clams - 128

The Slate Roof Bible - Index

clay slate - 21
cleats - 204, **243**
cleavage - 23
cleavage plane - 22, 23
collar ties - 195, 196
Colonial Works Co. - 275
comb - 219, 234
Conasauga - 135
Conowingo Dam (MD) - 119
continental drift - 25
contractors - 8, 11, 13, 14, 120-121, 189, 219
    gouging - 12
copper - 122, 182, 195, 204, 205, 240, **242,**
        243, 252, 262
    lead coated - 240, 242, 262
    sewer vent flashings - 261
    solder recommended for - 244
    sources of - **237**
    spouting - 220
    using - 243
    valleys - see "valleys, copper"
Copper Development Assn. - 242, 243
Copper Sales, Inc. - 237
Cornish - 143
Cornwall - see "England "
Cotswald stone - 36, 38
counter flashing - see "flashing, counter"
Countess - 56
co-workers - 157, 163
Crediton - 20
creosote - 266
cricket - 156
crinoids - 128
Crummette, David - 128, 130
Crystal Palace Exposition - 119, 125
curtain theory - 9
Cwt y Bugail Slate Quarries - 218
Czechoslovakia - 25

D

Dally Slate Company - 111, 218
Day, Arthur - 138, 140
dead load - 194
Declaration of Independence - 122
deep mine - 63
Delabole slate - 18, 48
Delaware River - 105
Delta (PA) - see "Pennsylvania"
Denninger Weather Vanes and Finials - 275
Design Services Inc. - 217
dew - 162
Diamond Run - 27, 102
diamond blade - 217
diamond saw - 94, 111
diamond wire machines - 144
Dinorwic Quarry (see "Wales")
ditch - 86
Docker, William - 122
dormer - 156
    flashing - 239
    step flashing - **261,** 262
    top flashing - **259**
Doubles - 56
Double-doubles - 57
drills - 85
drip edge - 178, **219,** 220
    metal - 201, 203
    repairing - 221
drip stones - 270

drumhouse - 60, 61
drywall - 217
Dublin, Ireland - 59
Duchess - 56
dump car - 89
Dutch lap pattern - 199, 200
dynamite - 85, 92

E

eaves - 156, 160, 178, 185
    curved - 277
electrical wires - 157, **162,** 177
electrocution - 163
electrolyte - 242
end slates - 203, 224
England - 18, 19, 20, 25, 31, 35, 36, 40, 119,
        125, 129, 187, 208, 209
    Aberdeenshire - 54
    Argyll - 54
    Construction fluctuations - 59
    Cornwall - 19, 48, 54, 72
    Devon - 20, 54
    Lake District - 31, 40, 54
    London World's Fair - 131
    National Trust - 72-73
    Tintagel - 19, 72
English - 140
epoxy adhesive - 217
Esco slate cutter - 172, 185
Europe - 8, 17, 33, 34, 121, 127
    Eastern - 92
Evergreen Slate Co. - 95, 96, 97, 185, 218

F

face-nailing - 12, 13, 224, 228
face (of a roof slate) - 200
Faill, Roger - 118
Fair Haven, VT - 43
Fairmount (GA) - 43
false cleavage - 23
fascia - 156, 202, 219, 220, 221
    hanger - 220
felt paper - 187, 192, 193, 202, 204, 205, 211,
    236
fiberglass roof membrane - 226, 232-233
fiberglass shingles - 9, 11
finials - 275
Finnish - 140
fireproof - 10, 12, 13
Fischer Artworks - 275
flashing - 5, 6, 156, 169, 204, 219,
        223, 229, **239**
    apron - 247, 248
    base - 245
    basic principles - 244
    brazed - 247
    cap - 245
    chimney - **244 - 250**
    counter - 245, 247, 248, 262
    dormer top - **257-258**
    side - 244-247
    step - 245, 260, **261,** 262
    top (on chimney) - 246, 248
        hidden leaks - 249
        on shed roof dormer - 257
    wedges - 245
floors - 217
flux - 244, 263, 264

fly (of ladder) - 175, 177
fossils - 25, 128
framing - 194, **195**
French method - 199, 200
France - 21, 25, 59, 173
frost - 162

G

gable end - 156
gable roof - 156
gable verge - 211
Galicia, Spain - 37
Galite - 134
galvanic action (corrosion) - 184, **242, 243**
galvanized metal, etching - 235
galvanized metal primer - 235
galvanized spouting - 220
galvanized steel - 240, **241,** 243
gambrel roof - 156, 166
Garbage Magazine - 9
geological forces - 21
Georgia - 16, 26, 43, 49, 127, **132-137**
    Atlanta - 132, 135
    Bartow County - 132, 135
    Blanceville Slate Mines - 134
    Cohutta Mtns. - 132
    Dug Down Mtns. - 132
    Fairmount - 71, 132, **135,** 136
    Fannin Co. - 132
    Geological Survey - 132
    Gordon Co. - 132, 135
    Murray Co. - 132, 135
    Pine Log Mtns. - 132
    Polk County - 132, 134
    Rockmart - 71, **132,** 134
        Presbyterian Church - 133
    Savannah - 134
    Silicoa Mountains - 132
    slate, exports - 59
        production - 134
    Van Wert - 133, 134
        Methodist Church - 133
Germany - 25, 59, 129, 174
Gibraltar - 25
Glendyne Quarry - 142, 144, 218
Gloucestershire - 36
Golden Gate Bridge - 135
gouge - 85, 86
graduated roof pattern - 199, 201
grain - 22, 23
Granville - see "New York"
gravestone - 28, 128, 133
Great Britain - 59
Great Depression - 131
Great Fire of London - 59
green lumber - 190, 191, 221
green slates - 15
ground ladder - 156, 163, **175**
    erecting - 176
    plumbing - 176
Grove City (PA) - see "Pennsylvania"
Gundlach Co. - 172, 185
gutters, built-in (box) - 221, 222, **278-279**
    replacing - 279
Gwydir Slate Quarry (Wales) - 53